The Development Of English Building Construction

he Development Of English Building Construction

BY

C. F. INNOCENT

Associate of the Royal Institute of British Architects
Honorary Lecturer in English Architecture at the University of Sheffield

New impression with new introduction
and bibliography by Sir Robert
de Z. Hall

DAVID & CHARLES REPRINTS

ISBN   0   7153   5299   7

First published in 1916 by Cambridge
University Press

This impression 1971 by David & Charles
(Publishers) Limited

Printed in Great Britain by
Clarke Doble & Brendon Limited   Plymouth

NEW INTRODUCTION

Charles Frederick Innocent was born in Sheffield in 1873 or early 1874, the elder son of a well-known local architect. He took up his father's profession and was elected as an Associate of the Royal Institute of British Architects in 1897. Four years later, a few months before the death of his father, he was taken into partnership. One would imagine that he was not too preoccupied with the need to follow his profession, from the quantity of reading, correspondence and travel disclosed in his book, and from his having completed the main body of his research within a dozen years. In addition, he was for a long period a member of the Council of the Sheffield Society of Architects and had antiquarian interests; and, save for a period during World War I, he was an honorary lecturer in English architecture to evening classes of the University of Sheffield from 1910–11 till 1923, when he died.

The climax of his career was the completion of the task of writing his book on the architectural history of the lesser English house, to fill a gap in architectural literature which, as he says in his preface, was absolute when he came to study the subject. The book was published in 1916 in the Cambridge Technical Series, and it is to be emphasised that *The Development of English Building Construction* was written for architects, though many archaeologists and antiquaries have since used the work. Issued in the middle of a war, the edition was small, only just over 700 copies, and it was never reprinted. Yet it has never been a forgotten book, and one modern writer after another has acknowledged his debt to a study which, though overtaken in part of its detail, has not, in over half a century, been replaced by a modern comprehensive work. One of those who used it in earlier years, the architectural historian J. H. Harvey, provides an apposite extract from Chaucer's 'The Parlement of Foules':

> For out of olde feldes, as men seith,
> Cometh al this newe corn fro yeere to yeere;
> And out of olde bokes, in good feith,
> Cometh al this newe science that men lere.

NEW INTRODUCTION

If one takes 1895 as about the time when Innocent became conscious of the lack to which he referred, there was in fact next to nothing in print about lesser secular architecture. Some of the local historians of the early part of the nineteenth century gave brief descriptions, particularly the north of England writers T. D. Whitaker and W. Thornber, who drew attention to the primitive construction of crucks.[1] But the tone for many decades was set by Sir Giles Scott, inviting his readers to 'look at the vernacular cottage-building of the day'.[2] With rare exceptions, the aesthetic approach dominated the examination of small houses till well into the twentieth century, though sufficient examination of plan and structure had taken place by the time that Innocent was writing, for him to give a respectable list of acknowledgements.

A fundamental work was, in fact, published in 1898, soon after Innocent had qualified as an architect, Addy's *The Evolution of the English House*.[3] Sidney Oldall Addy, a fellow citizen of Innocent, was a much older man and survived him by ten years. He was born in 1848, the son of a Derbyshire landowner and colliery proprietor, and after obtaining a degree at Oxford in 1871 was admitted as a solicitor in 1877. He was actively in practice in Sheffield, but the nature and length of his list of publications shows that he had ample leisure for antiquarian research. His principal early work was on the dialect of the neighbourhood of Sheffield, and there followed a stream of books and articles on almost anything that could attract an antiquary, from genealogy to folklore, from medieval cutlery to local history. It is from the position of a social historian that one must see his contribution to domestic architecture. Having identified the nature of primitive structures, with much about crucks, he passed to plan and social purpose; and the difference in principle between his work and that of Innocent is shown by its having been published in Swan Sonnenschein's Social England series. There was a good deal of continental comparison and development of theories, not nowadays acceptable, but the earlier part of the book gave it a wide appeal (to architectural students also) and after two reprintings it was revised and largely re-illustrated by J. N. (now Sir John) Summerson in 1933.

[1] Whitaker, T. D. *The History of Whalley* (1800–1, later 1806 and 1818, 4th ed, revised, 1876); *The History of Craven* (1805, later 1812 and 1878); Thornber, W. *An historical and descriptive account of Blackpool and its neighbourhood* (1837).

[2] Scott, Sir Giles, *Secular and domestic architecture* (1857).

[3] Bibliographical details in select bibliography.

Another early work to which Innocent made frequent references was Hughes' and North's *The Old Cottages of Snowdonia*, and this is to be singled out as the earliest work in which architects gave attention to structure, in this case almost entirely the cruck form, rather than style.[4] The aesthetic approach, however, dominated the period before Innocent's work was published. It was applied both to individual houses and to their setting in the landscape. A number of splendid books appeared, in which the contribution to knowledge of structure and plan was small, but which had the common factor of plenty of photographs and drawings. They laid the foundation for modern ideas and practice on conservation, but also provided much material for later students, not least by providing some record of buildings which no longer stand. From these works, some by eminent architects, developed a concept of regionalism based on geology, which has been well brought out by Sydney R. Jones, best known as an illustrator.[5]

It is convenient at this point to leave the chronological sequence and to mention, as an early climax to this type of work, those books on buildings in Batsford's British Heritage series. These were well written and well illustrated, and of particular relevance for the present purpose was *The English Cottage*, by H. Batsford and C. Fry.[6] Aesthetics and materials were still dominant, but the authors had a good deal to say about structure, paying tribute especially to Addy and Innocent.

Reverting to Innocent's period, little was being done by individuals about the systematic recording of buildings. Two institutions, however, came into being, whose investigators by degrees paid increasing attention to minor domestic architecture. These, in order of creation, were the Victoria History of the Counties of England, founded in 1899, the first of the 'topographical' volumes appearing in 1903; and the Royal Commissions established in 1908 on Historical Monuments in England and on Ancient Monuments in Wales and Monmouthshire. The amount of published work is so large as to preclude any particularisation, but it can be said, in general, that the Victoria County Histories had from the start a strong trend towards architectural recording and that the early

[4] Hughes, H. H. and North, H. L. *The old cottages of Snowdonia* (1908).

[5] Jones, Sydney R. *The Village homes of England* (The Studio, 1912); *English Village homes and Country Buildings* (1936, reprinted 1947) is similar, with some discussion of plan and structure.

[6] Batsford, H. and Fry, C. *The English cottage* (1938; later editions up to 1950).

volumes of the English Royal Commission, though they had little to say about minor buildings, made a significant contribution to the knowledge of the principles of timber building in the southern and eastern parts of England, which were unknown to Innocent. Some of the more recent volumes in each case are listed in the bibliography which follows.

These enterprises, however, could only provide summary records, covering the ground by degrees; and in the period between the two wars an ever-increasing number of private investigators set themselves to recording. Their results are mainly to be found in a variety of journals, largely those of the county archaeological societies, and prominent among those which gave space for the purpose were those of Birmingham, Kent, Surrey and Sussex. Selection into a short bibliography is impossible, and the remainder of this foreword must be related to works which had a wide impact, either as pioneers or as presenting a general view based on available research. (Bibliographical details will be found below, not in footnotes as hitherto.)

The first landmarks were in Wales. C. F. (later Sir Cyril) Fox, Director of the National Museum of Wales, wrote a slender book, published in 1932, *The Personality of Britain*, in which, utilising the technique of the distribution map, he demonstrated the division into two great archaeological zones, the Highland, with greater continuity, more backward, covering Wales and north, west and south-west England, and the Lowland, more advanced and with greater unity. From this developed the concept of cultural regions within a zone, and this was tested and applied between 1941 and 1948, while he was still director of the museum, in collaboration with Lord Raglan, in the rural parts of Monmouthshire. Close on 500 houses were studied, and the outcome, *Monmouthshire Houses*, from the medieval to the renaissance period, published between 1951 and 1954, is still, both for the theoretical approach and the manner of recording, a preferred textbook for many recorders. In the introduction, Fox's successor as director revived the term 'vernacular' for traditional architecture on the lesser scale, and this is used henceforth.

Preceding Fox by a few years, I. C. Peate, whose name is indissolubly linked with the Welsh Folk Museum and the Society for Folk Life Studies, had brought out in 1940 *The Welsh House*. The sub-title of the book was 'a survey in folk culture', and in his introduction, Peate expressed his distrust of the archaeological approach, especially the distribution map, and placed the construction of a house at the last, not the primary stage of investigation. The differing approaches have been salutary, though today the archaeologist, and more recently the architect, hold the field.

Already in other parts of Britain, other investigators who became prominent after the war had published their first studies, such as R. T. Mason, the first of whose articles in *Sussex Archaeological Collections* appeared in 1939, and W. A. Pantin, specialist on town houses, in *Oxoniensia* in 1937. But undoubtedly the leading private investigator of the time was J. Walton, like so many of his successors a teacher of scientific subjects. He never wrote a full-length book, and his interest was mainly directed to cruck buildings, on which he produced a series of booklets and articles, some direct recording, others on wider aspects, between 1947 and 1955, when his career had already taken him to Basutoland.

Meanwhile the post-war publications of the Royal Commissions demonstrated the increase of interest in farmhouse and cottage, and the same applied to the Victoria County Histories, notably those for Oxfordshire and Warwickshire. Of especial importance was the introduction of the study of vernacular architecture at the School of Architecture in Manchester by Professor R. A. Cordingley, not in a narrow technical sense but as a means to the study of folk life. Some account of the range of theses written by research students of the school appears in R. W. Brunskill's article 'A systematic procedure for recording English vernacular architecture'.

In the same period in 1952 was published (the research had been completed and the book written well before the war) an outstanding contribution to the history of building. This was the work of the historian Dr L. F. Salzman, working largely on public records and relevant medieval illustrations—*Building in England down to 1540*. Vernacular building is barely touched upon, but the principles are general, and in relation to the archaeology of the subject it is instructive to set Innocent's and Salzman's books together and to see how closely parallel can be the contributions of persons with completely different training and experience.

Sufficient names have been mentioned to demonstrate that there was a widely distributed interest in vernacular architecture in the post-war years, of the disciplined kind of which Innocent was the pioneer. But there was no forum. In 1952, N. T. Porter, a man without any formal training but civilised by many years of contact with teachers and students in Cambridge, and later settled in Shaftesbury, Dorset, conceived the idea of bringing some of the researchers together with the object of founding a permanent association. His qualification for taking the lead was an antiquary's knowledge of many buildings in England and Wales, fortified by a library, now in the possession of the Institute of Advanced Architecture of the University of York. He was actively in contact with T. W. French, an investigator of the Royal Commission on Historical Monuments (England) working in Dorset; Salzman was an intimate friend; Fox he had known since Cambridge days; and Walton's work was familiar. A meeting was arranged for purposes of discussion, and the most important conclusion was that the endeavour should be made to form a permanent organisation. This was brought about, under the name of the Vernacular Architecture Group, and an initial membership of about thirty emerged, under the leadership of Fox. Of the group, just under a quarter were official investigators (including a Victoria County History editor), a similar proportion curators and archaeologists, while the remainder were architects, architectural historians, historians and plain antiquaries, teachers being conspicuous among the last. From the outset the conception was that the organisation should be a group for discussion, never a formal society for exposition; and the principle of admission has been of persons qualified in their particular fields, and also of students and others who showed evidence of continued active interest in vernacular architecture. In spite of this limitation, membership has risen in nineteen years to close on 200, and the nature and proportions of membership are mainly similar, though supplemented by archivists, geographers and a wider range of those with official interests. An exception is that the proportion of informed amateurs steadily grows, and in this fact, and the increasing willingness of county and other societies to publish, lies the hope of extensive and rapid recording of buildings.

In recent years, a number of the universities have shown interest. There is an active research group, of staff and students, based on the Manchester School of Architecture. At York there have been summer schools; based on Birmingham there are extra-mural groups in Shropshire and Herefordshire; and conferences and courses have been organised elsewhere. Also, under the leadership of members of the Vernacular Architecture Group, there are local research societies in Essex, Surrey and Sussex.

Though this foreword has been limited to England and Wales,

interest in research does not cease there. VAG, as it is known internally, has had its Scottish and Irish members from early days, and some of its members are in active contact with fellow workers in Belgium, East Germany, the Netherlands, Sweden, and the United States of America, some of whom are members of the group. In South Africa, Walton, now permanently resident there, has been responsible for the foundation of a flourishing Vernacular Architecture Society.

The seeds planted by Innocent, with whom one must join Addy, are now perennial plants. Some of the fruits, in the form of published work, appear in the select bibliography which follows. This is restricted to some of the classics and to more general books and articles published within the past two decades, and selection can only be invidious, particularly when a number of the very numerous studies of individual buildings contain observations of general importance. The list could not have been compiled at all save for the painstaking enthusiasm of the Bibliographer of the Vernacular Architecture Group from 1954 to 1969, J. T. Smith.

BIBLIOGRAPHY

A select bibliography of writings on, or related to, the study of vernacular architecture in England and Wales.

GENERAL

Addy, S. O. *The evolution of the English house* (1898, reprints 1905 and 1910) New edition, edited by J. N. (Sir John) Summerson, in consultation with the author (1933)

Barley, M. W. *The English farmhouse and cottage* (1961, reprinted 1968) The scope of the book is well stated on the jacket: 'Into the place of the sentimentalist in search of the picturesque and the architect looking for a style, have stepped the local historian and the archaeologist.' Documents are extensively used. This is essentially the modern, disciplined successor to Addy's work, on a wider scale, and in the context of Highland and Lowland zones, and regions within them.

Barley, M. W. 'Rural housing in England, 1500–1640.' Chapter X in *The agrarian history of England and Wales*, vol 4, ed Joan Thirsk (1967) A skilful summary, extending to larger as well as vernacular houses.

Clifton-Taylor, A. *The pattern of English building* (1962) An exhaustive account of materials.

Foster, I. Ll., and Alcock, L. (eds) *Culture and Environment; essays in honour of Sir Cyril Fox* (1963) Among essays relevant to aspects of Sir Cyril's varied career, a substantial number are on vernacular architecture:

    Chap 13 Rigold, S. E., 'The distribution of the Wealden house.'
         15 Raglan, Lord. 'The origin of vernacular architecture.'
         16 Smith, J. T. 'The long-house in Monmouthshire: a re-appraisal.'
         17 Smith, P. 'The long-house and the laithe-house: a study of the house and byre homestead in Wales and the West Riding.'
         18 Peate, I. C. 'The Welsh long-house: a brief re-appraisal.'
         19 Pantin, W. A. 'Some medieval English town houses: a study in adaptation.'
         20 Barley, M. W. 'A glossary of names for rooms in houses of the sixteenth and seventeenth centuries.'

Fox, C. F. *The personality of Britain* (National Museum of Wales, Cardiff, 1932 and numerous editions up to 1959, especially the fourth revised, 1943)

Hoskins, W. G. 'The rebuilding of rural England, 1570–1640'. *Past and Present*, 4 (1953); reprinted as chapter 7 of *Provincial England*

BIBLIOGRAPHY

(1963) The first statement of a theme which has had a profound impact on the study of small secular buildings, and an illustration of the importance of the economic historian to studies of vernacular architecture. In his later works, chapter 9, 'Fieldwork: buildings', in *Local history in England* (1959) is the passage of most direct interest.

Osborne, A. L. *The Country Life pocket guide to English domestic architecture* (1967) A descriptive and illustrated glossary by an architect, the successor to his *Dictionary of English domestic architecture* (1954)

Peate, I. C. *The Welsh house: a study in folk culture* (1940; later editions, Liverpool, 1944 and 1946)

Salzman, L. F. *Building in England down to 1540* (1952, second edition, 1967)

Smith, J. T. 'The evolution of the English peasant house to the late seventeenth century: the evidence of buildings.' *J Brit Archaeol Ass*, 3s 33 (1970)

Smith, P. 'Welsh rural housing, 1500–1640.' Chapter XI in *The agrarian history of England and Wales*, vol 4, ed Joan Thirsk (1967) The first, and admirably clear, exposition in the modern idiom of the development of vernacular houses in Wales.

West, T. *The Timber-framed House in England* (Newton Abbot, 1971) A useful introduction for the general reader, extending to detail as well as structure and plan, and well illustrated. There is also concise advice on restoration and preservation.

Wood, M. E. *The English medieval house* (1965) A comprehensive work, bringing together the work of many investigators, including herself. The book relates mostly to larger buildings, but is invaluable for comparative purposes. There is an extensive bibliography.

RECORDING OF BUILDINGS

Brunskill, R. W. 'A systematic procedure for recording English vernacular architecture.' *Trans Ancient Monuments Soc*, NS 13 (1965–1966) An account of the system developed by Professor R. A. Cordingley in the School of Architecture of the University of Manchester, as a rapid and ordered means of dealing with the exterior of a house. The author is a lecturer in the school. Very numerous illustrations of types of materials and their use.

Brunskill, R. W. *Illustrated handbook of vernacular architecture* (1970) The author has long recognised the value of the informed amateur in placing buildings on record, and has participated in summer schools on the subject. This book is aimed at the non-professional enthusiast. Both materials and methods of construction are covered.

Council for British Archaeology. 'Notes on the investigation of smaller domestic buildings.' Research Report 3 (1956) (reprinted from *Archaeol Cambrensis*, 1955). Other articles on the approach to recording buildings are by Mason, R. T., in *The Illustrated Carpenter and Builder*, 146 (1957); Tonkin, J. W., in *Middlesex Local*

BIBLIOGRAPHY

Hist Bull, 12 and 14 (1961 and 1962); *Cornish Archaeol*, 4 (1965); and Wood, M. E., in *Trans Newbury Dist Fld Club*, 11, no 3 (1965).

Iredale, D. *Discovering this old house*, the revised title of *This Old House*, (Tring, 1968). The expansion of an earlier article in the *Amateur* (now *Local*) *Historian*, which has published several relevant contributions. The author is an archivist and most of the booklet is about types of records and the location of record offices, but there are useful notes for people entering on the subject, on plan, elevation and detail.

Antiquity has published a series of articles on archaeological recording, and of these one is of particular practical value to a student of vernacular architecture seeking to produce a record of the quality required for publication. The general title is 'Archaeological draughtsmanship: principles and practice', and the particular article 'Part III: Lines of communication', by Hope-Taylor, B., in vol 41 (1967).

REGIONAL AND LOCAL STUDIES

South-east England

Mason, R. T. *Framed buildings of the Weald* (1964, second, enlarged, edition, Horsham, 1969) The author takes an historical as well as a constructional approach. His study is based largely on his very numerous individual reports published over a period of nearly a quarter of a century, mainly in *Sussex Archaeol Collect* and *Sussex Notes and Queries*.

Parkin, E. W. The principal contributor to a series published in *Archaeol Cantiana* at intervals from 1962, 'The vanishing houses of Kent'.

Eastern England

Forrester, H. *The timber-framed houses of Essex* (Hertford, 1959) An examination of types and details, most of the examples having been drawn from the Royal Commission on Historic Monuments' (England) *Inventory of Historical Monuments in Essex*, 4 vols (1916–23)

Hewett, C. A. 'Timber-building in Essex.' *Trans Ancient Monuments Soc*, NS 9 (1961) Other articles on particular aspects in *Archaeol J*, 119 (1962), and *Medieval Archaeol*, 10 (1966).

Royal Commission on Historical Monuments (England) (referred to hereafter as RCHM). *County of Cambridge*, vol I, West Cambridgeshire (1968) The first inventory to give extensive treatment of vernacular rural buildings.

East Midlands

Webster, V. R. 'Cruck-framed buildings of Leicestershire.' *Trans Leicestershire Archaeol Hist Soc*, 30 (1954) Among the earliest of the catalogues which have provided background for later workers.

South Midlands

Fletcher, J. M. 'Crucks in the west Berkshire and Oxford region.'

BIBLIOGRAPHY

Oxoniensia, **33** (1968)

Rigold, S. E. 'The timber-framed buildings of Steventon (Berkshire) and their regional significance.' *Trans Newbury Dist Field Club*, 10 no 4 (1958)

Victoria County History. *Gloucestershire*, VI (1965) and VIII (1968) *Oxfordshire*, V, IX (1957–59)
Both histories in progress.

Wood-Jones, R. B. *Traditional domestic architecture in the Banbury region* (Manchester, 1963) The most important study hitherto of an English region.

West Midlands

Charles, F. W. B. Chapter on timber-framed buildings and comments *passim* in Pevsner, N., *The buildings of England, Worcestershire* (1968)

Jones, S. R. and Smith, J. T. 'The Wealden buildings of Warwickshire and their significance.' *Trans Proc Birmingham Archaeol Soc*, 79 (1960–1)

Tonkin, J. W. 'Ancient buildings, 1964.' *Woolhope Club Trans* (Herefordshire), 38, i (1964) The first of an annual series of notes, demonstrating the value of further recording, sometimes of new discoveries, in an area where the RCHM has made a survey directed mainly to larger buildings.

Victoria County History. *Warwickshire*, III–VI (1945–51). III includes Stratford-on-Avon. Later volumes, completing the county, relate to towns (see Town Buildings, below).

South-west England

Alcock, N. W. Articles mainly in *Trans Devon Ass*, from 104 (1962) and especially 110 and 111 (1968 and 1969), beginning a series entitled 'Devon farmhouses'.

Chesher, V. M. and F. J. *The Cornishman's house* (Truro, 1968). An historical approach to development from the earliest times till the eighteenth century, with full use of documentary material, and well supported by photographs and plans.

Jope, E. M. 'Cornish houses, 1400–1700.' In *Studies in building history*, ed E. M. Jope (1961)

RCHM *Dorset*, vol I, West Dorset (1952); II South-east Dorset, and III, Central Dorset (1970). Many photographs and plans are relevant to vernacular architecture. Vol IV, North and North-east Dorset, is in hand.

North-west England

Brunskill, R. W. 'The development of the small house in the Eden valley from 1650 to 1840.' *Trans Cumberland Westmorland Antiq Archaeol Soc*, 53 (1953)

Singleton, W. A. One of the pioneers from the Manchester University School of Architecture, whence came a number of research theses, many related to the north-western counties. Particulars are given in Brunskill, R. W., *Trans Ancient Monuments Soc*, NS 13 (1965–

BIBLIOGRAPHY

1966). A number of articles by Singleton on lesser domestic architecture in south-east Lancashire and north-east Cheshire were published in:

J Manchester Geogr Soc, 55 and 56 (1949–50 and 1950–2)
Cheshire Hist, 1 (1951)
Trans Hist Soc Lancashire Cheshire, 104 (1952)
Trans Lancashire Cheshire Antiq Soc, 65 (1955)

Victoria County History. *Lancashire*, III–VIII (1907–14). The area covered, before work was suspended, was the south and south-east of the county, and Lonsdale, and there is a fair amount of material with relevant interest.

North-east England

Raistrick, A. 'Dales building of the sixteenth and seventeenth centuries', *The Yorkshire Dalesman* (1941–2). Several articles, particularly in nos 7, 8 and 10.

Raistrick, A. *Old Yorkshire dales* (Newton Abbot, 1968) The place of village and farm in local history is discussed.

Stell, C. F. 'Pennine houses: an introduction.' *Folk Life*, 3 (1965) Based on a research thesis submitted to the Liverpool University School of Architecture. The area covered is that in the neighbourhood of Halifax, but there is material relevant to much of northern England.

Walton, J. *Homesteads of the Yorkshire dales* (Dalesman publication, Clapham, 1947)

Walton, J. 'Cruck framed buildings of Yorkshire.' *Yorkshire Archaeol J*, 37 iii (1950)

Walton, J. 'Early timbered buildings of the Huddersfield District' (Tolson Memorial Museum, Huddersfield, 1955)

Wales

Bevan-Evans, M. *Farmhouses and cottages: an introduction to vernacular architecture in Flintshire* (Flintshire Record Office, Hawarden, 1964)

Brooksby, H. 'The houses of Radnorshire.' *Trans Radnorshire Soc*, 38 (1968) and 39 (1969)

Fox, Sir Cyril and Raglan, Lord. *Monmouthshire houses* (National Museum of Wales, Cardiff). Part I Medieval (1951); II Sub-Medieval (1953); III Renaissance (1954)

Jones, S. R. and Smith, J. T. 'The houses of Breconshire.' Parts I–V, *Brycheiniog* (J. Brecknock Soc), 9–13 (1963–1968–9). In progress. The most exhaustive county survey so far made. Discussion includes social and economic background and fundamentals of house plan.

Royal Commission on Ancient Monuments. *Caernarvonshire*, I–III (1956–64) The first to give more than passing attention to vernacular architecture, though the inventory for Anglesey, 1937, has some references.

Smith, P. 'Welsh rural housing, 1500–1640.' Chapter XI in *The agrarian history of England and Wales*, vol 4 (1967) The first—and admirably clear—comprehensive view of Welsh vernacular architecture.

TOWN BUILDINGS

Forrester, H. *Timber-framed buildings in Hertford and Ware.* (Hertfordshire Local History Council, 1965.) Not profound, but an excellent basic gazetteer, with some descriptions, of about sixty houses.

O'Neill, B. H. St. J. 'Some seventeenth-century houses in Great Yarmouth.' *Archaeologia*, 95 (1953) Of wider importance than the title indicates.

Pantin, W. A. 'The development of domestic architecture in Oxford.' *Antiq J*, 27 (1947)

Pantin, W. A. 'Medieval inns.' In *Studies in building history*, ed E. M. Jope (1961)

Pantin, W. A. 'Medieval English town houses and plans.' *Medieval Archaeol*, 6–7 (1962–3)

Portman, D. *Exeter houses, 1400–1700* (1966) A comprehensive study.

RCHM *The City of Cambridge* (1959).

Victoria County History. *Warwickshire*, VII (1964) Contains a chapter on secular architecture, some of vernacular interest. *Warwick and Coventry*, VIII (1969) These are given as instances of recent work.

STRUCTURE

Charles, F. W. B. *Medieval cruck building and its derivatives.* (Soc for Medieval Archaeol, Monograph Series No 2, 1967) A study based mainly on Worcestershire buildings, but of wider importance, emphasising the problems of erection. Plentifully illustrated.

Cordingley, R. A. 'British historical roof-types and their members: a classification.' *Trans Ancient Monuments Soc*, NS 9 (1961) Definitions and very numerous drawings, with an introduction relating general types to the Highland and Lowland zones.

Fletcher, J. M., and Spokes, P. S. 'The origin and development of crown-post roofs.' *Medieval Archaeol*, 8 (1964)

Hewett, C. A. *The development of carpentry, 1200–1700: an Essex study.* (Newton Abbot, 1969) The range is wider than the sub-title indicates. The manner of erecting buildings is well described and illustrated. Among techniques, jointing has particular attention. Decoration, including mouldings, is also covered.

Smith, J. T. 'Medieval aisled halls.' *Archaeol J*, 112 (1955) A modern approach, against the background of earlier studies.

Smith, J. T. 'Medieval roofs: a classification.' *Archaeol J*, 115 (1958) Based on the same principles of approach as the previous entry, and, though going beyond the range of vernacular architecture, fundamental to its study. Important distribution maps.

Smith, J. T. 'Cruck construction: a survey of the problems.' *Medieval Archaeol*, 8 (1964) Part of the subject matter of the previous entry developed in depth, and in relation to continental material.

Smith, J. T. 'Timber-building in England.' *Archaeol J*, 122 (1965)

BIBLIOGRAPHY

The series of studies is here extended from roofs to walls. After an introductory critique of earlier publications, the theme is developed with the aid of illustrations and distribution maps, of three carpentry traditions, the eastern, western and northern.

All the works of this author are plentifully provided with references, many of which are briefly discussed.

DOCUMENTARY SOURCES

The field is so vast that only the briefest general introduction can be given, here and in the succeeding section.

Emmison, F. G. *Archives and local history* (1966) An introduction to classes of local records, and to repositories.

West, J. *Village records* (1962, revised edition 1966) Based on Worcestershire archives. A section on inventories and a list of printed inventories in other counties. For some of these see Emmison, F. G., 'Jacobean household inventories' (*Bedfordshire Hist Rec Soc*, 1966); Havinden, M. A., *Household and farm inventories in Oxfordshire* (1966); Steer, F. W., *Farm and cottage inventories of mid-Essex, 1635–1749* (1950, revised edition Chichester, 1970)

AGRARIAN BACKGROUND

Aslen, M. S. *Catalogue of printed books on agriculture published between 1471 and 1700* (Rothampstead Experimental·Station and Library, second edition, 1940)

Finberg, H. P. (general editor) *The Agrarian History of England and Wales*. Vol IV, 1500–1640, ed Thirsk, J. (1967)

Hilton, R. H. 'The content and sources of English agricultural history before 1500.' *Agric Hist Rev*, 3 (1955).

Marshall, W. *The review and abstracts of the county reports to the Board of Agriculture*, 5 vols (1818, reprinted Newton Abbot, 1968)

Tate, W. E. *The English village community and the enclosure movement* (1967) Many bibliographical references and critical reviews of recent works. In an appendix a full list of the late eighteenth/early nineteenth-century series *A General View of the Agriculture of . . . shire* (ie the reports reviewed and abstracted by Marshall, W., above).

Thirsk, J. 'The content and sources of English agricultural history after 1500.' *Agric Hist Rev*, 3 (1955).

CONSERVATION OF BUILDINGS

Edmunds, R. C. *Your country cottage: a guide to purchase and restoration* (Newton Abbot, 1970) A practical approach.

Harvey, J. H. 'Conservation of old buildings: a select bibliography.' *Trans Ancient Monuments Soc*, NS 16 (1968–9) Over 200 items, classified, being a revised version of the list which appeared in the same journal, NS 6 (1958)

PREFACE

WHEN the writer was studying the development of English architecture for the professional examinations, many years ago, he found that there was hardly any information easily available as to the design and construction of the smaller secular buildings. The books which described English mediæval architecture were almost exclusively filled with descriptions of great ecclesiastical buildings, and, similarly, the authors who wrote of the later architecture seemed hardly to descend below the houses of the lesser gentry. But the splendid cathedrals and abbeys of the Middle Ages, with their stone walls faced with carefully prepared ashlar, were as superior to the ordinary buildings of their time as are modern town halls and municipal buildings to the houses of the ratepayers from whose contributions they are built and maintained; they occupied a position above and apart, and when the unimportant buildings are investigated, it appears as though a descent has been made into some under-world of strange materials and curious methods. This however must not be carried too far, for the *Records of the Borough of Leicester* show that the saltpetre maker, whose work will be described later, was permitted in the year 1589 to throw down all the mud walls 'apt for his purpose in the late suppressed abbeys and houses of religion, and old castles.'

The use of such materials as thatch and mud and wattle is prevented to-day by the building by-laws, which are useful indicators of the attitude of the governing class towards building. The unimportant buildings have rarely been restored, and in that have a value for the artist and the craftsman which those of more importance no longer possess, and although such humble old buildings may be usefully studied for lessons in line and form and colour, we are here more concerned with their historic relationship to the building construction of the present time. But this is too large a subject to be treated *in extenso*, and the less known forms have therefore been described more fully. The crucks have received more attention than the posts and trusses, and thatch has been preferred before slates.

The available materials are manifold though often fragmentary, and of those which the writer has collected, only such have been used as seemed to be essential to the subject. The necessary size of the book must be responsible if in some cases the statements seem to be too summary, and the argument occasionally to advance *per saltum*.

The highways and byways in which the materials have been sought, are indicated in the first chapter, and, later, the place of origin of each statement is named in the text or in a footnote, as this was felt to be desirable with a new subject. In succeeding chapters the most primitive types of building still to be found are examined and the development of the principal framework from attempts adequately to support the roof ridge is described. After the woodwork, the carpenter or wright, his tools, and his materials naturally receive attention. The chapters which follow deal with the walls, floors, and roof coverings, and openings in the building, as doors and windows. Of these, as of the framework, an attempt has been made to trace their evolution to the forms which are in use to-day, and in the final chapter the principal modern materials, whose history is only short, and the apparent tendencies of the time are discussed.

This book has been extended from a series of articles by the writer which appeared in *The Building News* during 1912 and 1913, and the writer's thanks are due to the editor for permission to republish them in this form. The writer's hearty thanks are also due to all who have given him information, especially to Mr S. O. Addy, M.A., author of *The Evolution of the English House*, who supplied much information which would not otherwise have

PREFACE

been obtained, to Mr Joseph Kenworthy, whose knowledge of South Yorkshire buildings has been most useful, and to Mr Thomas Winder, A.M.I.C.E., whose *Half Timber Buildings in Hallamshire* first showed the unexplored interest of the construction of our minor buildings. Of students on the Continent the writer must name Mr Bernhard Olsen, of the Danish Folk Museum, at Copenhagen, and also the late Herr K. Rhamm, author of various books which treat, *inter alia*, of the building construction of the Teutonic and Slav countries, and who freely placed his wide knowledge at the author's service.

Cordial thanks are also due to those persons who have allowed an inspection of their houses and farm buildings, a visit which was usually unexpected and often at inconvenient times.

The majority of the illustrations are from drawings and photographs by the author but grateful acknowledgement is made to The Cumberland and Westmorland Archæological Society per Mr W. G. Collingwood, M.A., F.S.A., and Mr H. S. Cowper, F.S.A., Mr Bernhard Olsen, Herren Vieweg und Sohn, Messrs George Allen and Co., Ltd, Mr John D. Watson per Mr D. J. Roberts, A.R.I.B.A., Mr Joseph Kenworthy, The Woolhope Field Club and Mr Robert Clarke, Mr H. H. Hughes, A.R.I.B.A., and Mr H. L. North, F.R.I.B.A., The Cambrian Archæological Association per Rev. Canon Morris, The Proprietors of *Country Life*.

C. F. I.

April 1916.

CONTENTS

CONTENTS

LIST OF ILLUSTRATIONS

LIST OF ILLUSTRATIONS

LIST OF ILLUSTRATIONS

CHAPTER I

INTRODUCTORY

Sources of information—Old building accounts—Books of agriculture and topography—Glossaries and dictionaries—Buildings and their occupiers—Preservation of buildings.

The materials for the study of old English building construction are varied and generally fragmentary.

Much may be learnt from the accounts of the expenses of the old abbeys and corporations, which have been published by societies like the Surtees Society, and by such municipalities as the Corporation of Leicester. Although these accounts are generally concerned with buildings of a more important kind than are described in this book, it must be remembered that the methods of construction filtered down in the course of time from one stratum of society to another. Of this the external whitening of buildings is an example: a Norman abbot of St Albans caused the abbey church to be whitewashed, and boasted that it shone like snow, and Westminster Hall was whitewashed for Edward I: gradually the whitewashing of buildings descended in the social scale, until in the nineteenth century only the most humble buildings, such as cottages, were externally whitewashed. The renewal of its use for more important buildings in the last century was due to the so-called æsthetic movement.

It is thus evident that we must understand, to some extent, the more important and expensive buildings of the past in order to learn the descent of the ordinary construction of the present day.

The study is complicated, because old and humble buildings, such as cottages, may themselves have descended in the social scale in the course of their existence, like the good old furniture which is often found within them. Mr C. E. Clayton has found

this to be so in Sussex, and says that the oldest cottages 'certainly
do not represent the dwellings of the labouring class of the period
of their erection[1],' and the writer has found similar conditions in
the Sheffield district: a cottage at Treeton, South Yorkshire, now
occupied by a worker in a coal-mine, is said to have been the
former manor-house of the village. The old palace of Hatfield,
which became the stables of the early Renaissance mansion, is an
example on a greater scale.

It is interesting to observe the persistence with which forms
and methods of construction continued in use in this country,
century after century, and R. Henning noticed the same con-
servatism in the construction of German houses. He writes of
the strange persistence with which the characteristics extend
(*hineinragen*) from the earliest times up to the present day[2].

Monsieur C. Enlart in his *Manuel d'Archéologie Française* has
remarked upon the same archaism in French buildings. He says
that several pretty Gothic flèches in stone, which exist near Amiens
and Abbeville, only date from the seventeenth or even the
eighteenth century, and that in Brittany the seventeenth century
architecture is in the style of the early Renaissance, which would
indicate that at the earlier period the Breton style remained
Gothic[3].

Humble buildings, therefore, cannot be accurately ranged in
chronological manner, but they have, rather, to be considered in the
order of their development from the simple to the complex. Their
chronology has to be reckoned by types and not by centuries, and
the words 'old' and 'early,' so frequently used in this book, are thus
somewhat removed from their usual meaning.

Other sources of information are the books on agriculture
published about a century ago, when the ancient system of farming
in this country was being superseded by improved methods: they
often contain descriptions of buildings then in existence and
constructed after the traditional methods. Such, for instance, are
the works of William Marshall, an agricultural topographer of the
eighteenth century, and the contemporary series of volumes of
surveys of the agriculture of the counties published by the Government.

---

[1] 'Cottage Architecture' in *Memorials of Old Sussex*, p. 288.
[2] *Deutsche Haus*, p. 163.
[3] *Manuel d'Archéologie Française*, I, p. 84.

Useful material may also be found in topographical works, and the accounts of travellers in the more remote parts of these islands, when journeys in this country were looked upon as serious undertakings worthy of record. Earlier than these are the mediæval surveys of the possessions of the great nobles and the religious establishments, such as the *Liber Henrici de Soliaco, Abbatis Glaston* in the year 1189, and they also are of value.

Three great modern dictionaries, the *English Dictionary on Historical Principles* published by the University of Oxford, the *Dictionary of Architecture* published by the Architectural Publication Society, and Professor Wright's *English Dialect Dictionary*, also provide the student of old English building with rich material.

Mediæval glossaries, such as the *Promptorium Parvulorum* and the *Catholicon Anglicum*, and English glosses on Latin texts, such as those published in the Wright-Wülcker *Vocabularies*, are also useful, as they give the early names of the parts of ordinary buildings, and indicate the construction almost as much by their omissions as by their inclusions; although the glossaries and dictionaries are somewhat dull reading, they provide valuable information.

Technical books for the workers in the building trades were first produced in the seventeenth century; at first they supplemented the traditional methods of oral instruction in the crafts and have now almost supplanted them. As far as the writer knows, the first technical series was published by Joseph Moxon, under the title of *Mechanick Exercises or The Doctrine of Handy Works* in the year 1677, and it is frequently quoted in the following pages. Some of his instructions are curiously modern[1]. His *Exercises* included trades other than building, such as printing, and ran through several editions. Since Moxon's time the books on building construction have been produced in ever-increasing numbers.

Secular buildings of small importance have received much more attention in the Teutonic countries of the Continent than has been given to them in these islands, and there is a very copious literature on the subject, such as K. Rhamm's *Urzeitliche Bauernhofe* and his *Altslawische Wohnung*, Mejborg's *Gamle Danske Hjem,* and

---

[1] Such as the warning that workmen, in hot weather, do not like to dip every brick in a 'pale' of water, as they should, because it is troublesome and makes their fingers sore, preferring to throw 'pales' of water on the wall after the bricks are laid.

Gudmundsson's *Privatboligen paa Island i Sagatiden.* These
foreign books, which tell of the construction of the old ordinary
buildings, not of the castles and cathedrals, in the lands from which
the English nation is supposed to have come, enable comparisons
to be made with our English construction, and help to some
conclusion as to its racial origin. The evidence shows, broadly,
that although our architecture reached us by various routes from
the Mediterranean lands, our building construction is of Northern
origin.

The references in the text show the wealth of literary material
which is available in obscure places, but the buildings themselves
are more valuable than modern descriptions or contemporary
statements of expenditure. The old cottage, or farm building,
is abnormal which does not illustrate some obscure old description
or some difficulty in development. If the building is ruinous or
in process of destruction so much the better, for then the construction lies open to examination. An enormous destruction of the
minor buildings has taken place recently, and is still in progress.
In some villages no old buildings are left. As an example,
Mr W. F. Price, writing of the two varieties of Lancashire cottages,
those of 'clam, stave, and daub' with thatched, and those of stone
or brick with flagged or thatched, roofs, says: 'Both types are
rapidly disappearing, and are being gradually replaced by the
uninteresting and featureless dwellings of the present day. From
the aesthetic point of view the loss of these quaint old buildings
is irreparable, for one can hardly call to mind a typical bit of
Lancashire landscape where the elimination of the little thatched
and whitewashed cottage would not be distinctly felt. This old
cottage architecture is picturesque and homely: there is no effort
in construction, no frivolous and unmeaning detail introduced to
mar its dignity, and the forms and colour are always pleasing
and restful[1].'

The reason for this wholesale rebuilding of the cottages lies in
a rise in 'culture stage' which is taking place with the artisan or
working class, analogous to that which took place with the middle
or yeoman class in the seventeenth century, and which then brought
about the destruction of almost all the old houses and the adoption
of a more advanced type of middle-class house in the North of
England. It is to be hoped that before the destruction of cottages

---

[1] 'Homes of the Yeomen and Peasantry' in *Memorials of Old Lancashire*, I, p. 255.

proceeds much further, those of the various districts of England may be described, in the same manner as has been done for one district by Messrs Hughes and North in their excellent *Old Cottages of Snowdonia.*

William Morris said that the homely old English cottages were models of architecture in their way. They have been called 'unarchitectural' but all's fair that's fit, and they are valuable as examples of the appropriate use of materials, as illustrations of fitness to site and surroundings, and as specimens of architectural development, for just as the finest man had his origin in a simple cell, so the finest examples of our architecture can be traced back in their origin, step by step, to simple 'unarchitectural' buildings.

The old farms and old cottages scattered up and down the land possess an essential requirement of successful building in that they appear to be part and parcel of the landscape. As Emerson wrote of certain more important buildings, 'These buildings grew as grows the grass.' The tile hanging of Kent, the brick nogging of Hertfordshire, the timber and plaster of the Western Midlands, and the stone walls and stone slates of 'the backbone of England' all seem to be as much a part of the landscape as the hedges and the trees, and yet it is unlikely that the beauty of these buildings was apparent to their builders. Their beauty came as the spontaneous product of the hands of their constructors, who were in a stage of culture in which technical ability produces works of art naturally and unconsciously, unlike the technically skilled workers of present day civilisation, who can only design 'works of art' after much training. Our smaller buildings became ugly when technique had become highly skilled and the workers had mastered their materials completely: this took place gradually and was accomplished about the commencement of the present century. In this movement to a higher culture stage, for such it is, free and compulsory education for all has played no inconsiderable part, and the older people, the survivors of the old order which is passing away, are another valuable source of information as to the old methods of construction. Much may be learned by conversation with them, for in their young days building was not so highly specialised as it is now, and it may sometimes be found that they have played their part in the erection of the cottages in which they live.

Not only have the old ordinary buildings received more attention in the literature of the Scandinavian countries than they have with us, but greater efforts have been made to preserve the buildings themselves, with all their fittings and furniture. As wood is the usual building material, it has been possible to transport the buildings to open-air museums, of which the most important are at Skansen, near Stockholm, in Sweden, and Lyngby, near Copenhagen, in Denmark. The former was founded by Dr Hazelius and the latter has been the life-work of Mr Bernhard Olsen. The old cottages and farmhouses there, as Mr Arthur Hayden has said, 'have their obsolete agricultural implements, their old methods of fencing, and quaint styles of storage. The furniture stands in these specimen homes exactly as if they were occupied. It is a remarkable open-air museum, and the idea is worthy of serious consideration in this country. Old cottages and farmhouses are fast disappearing and the preservation of these beauties of village and country life should appeal to all lovers of national monuments[1].'

---

[1] *Chats on Cottage and Farmhouse Furniture*, p. 11.

CHAPTER II

PRIMITIVE FORMS OF BUILDING

Flakes—Conical huts of wood and turf in England and on the Continent of Europe—Oblong huts—The ridge pole—Its history and its influence in later constructions.

The beginnings of building are to be found in temporary screens of brushwood piled up as a protection against wind, weather and wild animals. To-day such screens are used by civilised men when hunting, and they form the substitute for homes among savages in a low stage of culture, such as those in the Kalahiri desert, and in the interior of Australia.

A little progress has been made when the brushwood has been woven round, and in-and-out of, stakes, and this is the most primitive form of building of which English records remain—if the term building may be applied to such simple and elementary beginnings. In the year 1511, the *Records of the Borough of Nottingham* show that the Corporation paid for ' ii fleykes to be set bytween ye masons and ye wynde.' The word flake is still in use in the dialects ; as an example, it is defined in Baker's *North-amptonshire Glossary*, as ' formed of unpeeled hazel or other flexible underwood closely interwoven or wattled together between stakes, like basket-work.' Further progress has been made to-wards a roof, when the woven brushwood or wattlework is bent over at the top, and the natural result of this is a similarity between wall, roof and ceiling in simple and early constructions. In ' Le Cantonnier,' a chalk drawing by J. F. Millet, the French artist, the countryman is sitting under a tall and bent flake or hurdle, which is propped by a leaning pole.

The next advance was to arrange the stakes or poles and the in-woven wattlework so that a space was both covered and enclosed, and there is a German theory that this was done, at first, to protect the domestic fire, and that it was a woman's invention. The conical huts are the most primitive of the space-enclosing and covering buildings. Conical huts of branches fixed in the ground and tied together at their upper ends were used fifty years ago by agricultural labourers during harvest, and also by shepherds and goatherds : if required to be more than merely temporary, they were covered with earth, boughs, or straw[1].

At the present time, the best-known of these buildings are the wigwam-shaped huts of charcoal burners, which are in widespread use in Europe. The kind of hut in use in the neighbourhood of Sheffield has been described by Mr Thomas Winder as follows : ' They are built of a number of thin poles laid together in the form of a cone; the feet are placed about 9 in. apart, and they are interlaced with brushwood. A doorway is formed by laying a lintel from fork to fork, and the whole is covered with sods laid with the grass towards the inside, so that the soil may not fall from them into the hut. A " lair " of grass and brushwood is formed upon one side and a fire, often of charcoal, is lighted upon the hearth in the threshold[2].'

In a handbook to *The Historical and Ethnographical Department of Skansen*, the author, Mr Axel Nilsson, after describing rectangular huts, states that Mr Winder's ' description almost exactly tallies with what we know of the cone-shaped charcoal-burners' huts used in other parts of Sweden.'

In South Yorkshire the writer has found a charcoal-burners' hut of a somewhat less primitive type than that described by Mr Winder. Three poles, erected in the form of a tripod, formed the principal framework : they were not tied or fastened together in any way at their summits, but held by the notches of their twigs, and their lower ends rested on the bare ground. On these three poles, acting as a frame, other poles were laid, close together, overlapping somewhat at their apices, and forming a cone of poles of a diameter of 9 ft. on the ground at the base. The height inside was 6 ft., which was just high enough for a man to stand upright in the centre of the hut : the measurement on the slope outside is 9 ft., so that the elevation of the hut is, roughly, that of an equilateral triangle. A covering of sods was laid on the poles, with the grass downwards and the whole pressed together, so that no rain could get through. The conical kitchen-huts (*kok-skalen*), of

---

[1] *Dictionary of the Architectural Publication Society, s.v.* Hut.
[2] *Builders' Journal,* III (1896), p. 25.

the Swedish summer farms, are very similar in construction : they are illustrated in the guide to Skansen.

Fig. 1 shows this charcoal-burners' hut and the wood pile, prepared for burning, in the wood ; and a view of the hut to a larger scale is given in Fig. 2.

Mr H. S. Cowper has described[1] a somewhat similar hut, but of a more advanced type in that the tops of the posts forming the tripod were fastened together by a withy, and the sods were laid overlapping each other like tiles. Poles, etc., were laid against the

Fig. 1.   South Yorkshire charcoal-burners' hut of branches and turf (to left)
and wood pile prepared for burning (to right).

sloping roof-wall of the hut in order to prevent these sod-tiles from being loosened by wind or rain. The hut was 7 ft. 9 in. high inside, and its width was 11 ft. Mr Cowper thinks that these circular huts represent the 'woodland wigwams of the Britons of Ancient Cumbria,' but it is possible that similar huts were used as dwellings until much later times, and Scandinavian influence is suggested by the poles laid on the sods, and by the sods themselves. It is curious that the 'houses' of flat stones laid on the

<hr />

[1] *Transactions of the Cumberland and Westmorland Antiquarian and Archæological Society*, XVI (1901).

ground, which children in South Yorkshire make for themselves in play, are always roughly circular in shape.

In an article on the forest of Compiègne in the *Century Magazine*, September, 1912, there is an illustration of a wood-cutters' hut in the forest: it is similar in appearance to the hut shown in Figs. 1 and 2. Such huts are said to be 'made so warm by their blankets of turf, that whole families may live comfortably in them for a couple of seasons. Chickens scratch before the door,

Fig. 2. Charcoal-burners' hut to a larger scale.

women do their washing on the paths, and children play among the leaves.' A German example, of similar construction, from the neighbourhood of Ruhla, has been illustrated by Moritz Heyne: the poles which form the foundation of the fabric are four, and not three, in number, and have forked ends, the natural branching of the tree, fitted into each other[1]. These conical huts only vary within narrow limits, and their widespread use in woodlands in North-Western Europe seems to imply that they are the descendants of the houses of some ancient forest-dwelling people. It is unlikely that there has been any intercourse

[1] *Deutsche Wohnungswesen*, p. 23.

between the charcoal burners of this country and those of Sweden, or between them and the woodcutters of France, and the type has probably spread from some common centre, and represents a primitive type of dwelling which has remained in use without further development, owing to the restriction of its employment and the simple requirements of its users. In the face of the European distribution it is curious that such conical huts are held to be peculiar to peoples living in open country.

When the circular hut was large, a centre pole was used as a support. In North Wales the huts are called ' Irishmen's cots,' and are as much as 24 ft. in diameter, and some contain a centre stone on which the post is supposed to have stood. At the Glastonbury lake-village the stumps of the central oak posts have been found in position.

The shape of the round house must have made it inconvenient and its size was limited by the mode of construction, which did not admit of ready enlargement. The next development of the house was to make it oblong in plan, straight-sided with rounded ends, and the builders were at once faced with a difficulty in supporting the tops of the poles which formed the framework of the straight sides of the building. Their methods with the circular huts have been described : in the oblong house they overcame this difficulty of a support for the slanting poles by the use of a horizontal pole, against which the poles might be leaned or to which they might be fastened.

The construction of wooden roofs has never progressed beyond this stage, and the majority of modern roofs are formed of rafters leaned against a ridge-piece. The rôle of the ridge-piece in gabled buildings is merely that of a convenience in the fixing and fastening of the rafters, but this has only been understood in recent times, and as the old builders believed that the ridge-tree bore the weight of the roof, they endeavoured to make it sufficiently large and strong to carry that weight, and they took pains to give it adequate supports ; these are the cardinal points in the history of our English building construction, for upon them everything in the main framework or carcase has depended and from them have descended our modern methods of building. The ridge-piece, the highest part of the roof and now insignificant, together with its supports, are the keys to the development of the construction of the building during many centuries.

The most primitive form of English building with a ridge-piece known to the writer was a form of charcoal-burners' hut formerly used in South Yorkshire, but not now constructed; it has been described as formed of two tripods, with the tops of the tripods connected by a ridge-pole, and with the door in the side. It was evidently formed by setting up the framework of two huts side by side, and then completing the construction to form one hut. This type is intermediate between the ordinary conical hut and that of the bark-peelers of High Furness which Mr Cowper has described

Fig. 3.   A bark-peelers' hut in High Furness.   The shape is oblong and the door is in the side.   The walls are of wattle packed between with earth and the roof is covered with sods, held down by poles.

in the paper which has already been quoted. Of this bark-peelers' hut he writes : ' Instead of three, four strong poles are selected, and the tops being lashed to a short ridge-pole 4 ft. long, the four feet are planted on the ground at the four angles of a parallelogram of about 13 ft. by 8 ft. Side walls with rounded corners, and con-structed of two faces of wattle packed between with earth, are then raised to a height of 2 ft. On the top of this wall, lighter poles of elder, birch and ash are then placed together, with their top ends supported against the ridge-pole. The sodding is then proceeded with as in the colliers' huts, but it only extends down to

the top of the wattle wall.' On one side is the door, and opposite
is a stone-built hearth, with a specially constructed chimney. The
'hut measures internally 13½ ft. by 8½ ft., and is 10 ft high. It is
for four persons, and is much roomier and more comfortable than
the colliers' hut,' which is only inhabited for a month or two, and,
as the colliers work day and night, they get their food from the
farms, and have neither cooking-place nor chimney. The bark-
peelers' huts are intended to stand longer, the occupiers have more
leisure and the construction is of a more advanced type. The
illustration (Fig. 3) shows a hut of the High Furness type.

Such structures as the bark-peelers' huts have been called sod-
houses, but, strictly, a sod-house is one whose walls are constructed
of sods in horizontal layers. The circular and oblong huts show
the origin of the ridged roof, but historically they lead into a cul-
de-sac, as they do not seem to have influenced later developments
in construction.

The ridge-piece has a very high antiquity, and roofs without
ridge-pieces are a later development[1]. There is documentary
evidence for ridge-pieces in Saxon England as early as the eighth
century. In Old and Middle English, the ridge-piece was called
'first,' and the word occurs many times in the Wright-Wülcker
Vocabularies as a translation of 'laquearea' and 'festum,' from the
eighth to the fifteenth century: in a glossary of the latter century
it is said to be the timber at which all the rafters come together
('est lignum ad quod omnia tigna conveniunt'). Much earlier
than this, in the tenth century, the author Byrhtferth described
a man building a house as fastening the rafters to the first ('tha
raeftras to thame fyrst gefaestneth[2]'). A thirteenth-century curse
threatened that 'the rof and the virste shal ligge on thine chynne[3].'
In two fifteenth-century English-Latin Dictionaries, the *Catholicon
Anglicum* and the *Promptorium Parvulorum*, the word 'rufe-tre'
is translated by 'festum,' and in the former dictionary by 'doma,'
in addition. The *English Dialect Dictionary* shows the various
names of the ridge-piece or pole in the English dialects. It is
called 'roof-tree' in Northumberland, 'ridge-pole' or 'ridge-pow'
in Cheshire, 'rig-tree' in West Yorkshire, 'ridge-tree' in the
Western Midlands, 'rig-baulk' in Lincolnshire and Leicestershire,

[1] This was clearly shown by K. Rhamm in *Urzeitliche Bauernhofe*.
[2] *Anglia*, VIII, p. 324.
[3] *Oxford English Dictionary*, s.v. First.

'rig-piece' in Leicestershire and Northamptonshire, and 'rig-tree' in the counties of York, Lincoln, Rutland, Leicester and Northampton. Moxon called the ridge-tree 'the first of the house[1],' and the word survives with the old meaning in the Western Midlands. At the present time 'first-piece' is used for 'ridge-piece' in Cheshire, and 'first-pole' in Shropshire.

Vitruvius, the Roman writer on architecture and building, advised the use of a ridge-piece ('columen') in roofs of both large and small span. Ridge-pieces are mentioned in the ancient Laws of Wales, and in the Triads, under the name of 'nen-pren,' in which 'pren' is a piece of wood and 'nen' the sky or roof. The writer has found by personal enquiry among the old people in North Wales that the term still survives. This high antiquity of the term shows that the Celtic buildings had ridge-pieces before that intrusion of English methods of building into Wales, which was brought about by the so-called Latin monks in the North and by the Anglo-Norman conquest in the South.

Herr K. Rhamm showed that Scandinavian buildings had ridge-pieces in early times, as they had also in the countries of High German speech. Between Scandinavia and the High German countries lie the Low German speechlands, where Herr Rhamm considered that the form of roof called 'sparrendach,' in which there is no ridge-piece, was used from the earliest times[2]. Herr Rhamm's knowledge of old building construction-methods in Central Europe was very great and his conclusions cannot be disregarded. As the early Teutonic invaders of England came from the countries where the 'sparrendach' reigned, it is possible that they adopted the ridge-piece from Romano-British buildings.

[1] *Mechanick Exercises*, Carpentry, Plate 11.
[2] *Urzeitliche Bauernhofe*, pp. 254 et seq., 539 et seq., etc.

CHAPTER III

PRIMITIVE FORMS OF BUILDING (*Continued*)

Upright forked poles—Their world-wide use in building—Furcae of gallows, etc.—Buildings rectangular in plan—Forks in old German and Scandinavian buildings—Variety of forked construction in Jutland.

The most simple method of supporting the ridge-pole is that in which the pole is sufficiently long and pliable to allow of the ends being bent downwards to the ground and the pole is, roughly, arch-shaped : this is still practised by savage peoples, but there is no evidence for its use in this country.

When the pole cannot be bent, the most simple and usual method of carrying it horizontally is to support it by an upright pole under each end, and such poles generally have forked ends in which the ridge-pole can lie without danger of side-slip. The use of such forked poles is widespread in savage and in semi-civilised lands, and extends to remote places in the Western Pacific Ocean. Mr R. W. Williamson has illustrated their use by *The Mafalu Mountain People of British New Guinea*[1] in the buildings called 'emone,' and Lieut. Boyle T. Somerville has described the erection of such a building in Malekula, New Hebrides, as follows : 'Two stout uprights cut to the required height, with a natural fork in the head of each, were driven into the ground at the desired distance apart, and a ridge-pole, well trimmed and as straight as could be procured, was firmly lashed on the forks with pandanus fibre and strips of bamboo.' The rafters were then formed by bamboos bent over the ridge-pole[2].

References to such forked posts occur in a Welsh Triad, which is translated by Professor J. E. Lloyd, the historian of early Wales, as follows : 'Three necessaries of the cot of a summer-bothy man : a roof-tree (nenpren), roof-forks (sing. nennforch, pl. nenffyrch) and wattling : these he may cut in any wild-growing wood he pleases.'

---

[1] In his book having that title.
[2] *Journal of the Anthropological Institute*, XXIII (1894), p. 372.

Professor Lloyd has kindly informed the writer that the manuscript of these Triads cannot be traced much farther back than the year 1685, but in the 'hafod,' or summer-bothy, the composer 'seems to have been describing something which was both ancient and well known to him.' The 'hafod' was occupied when the cattle were feeding on the uplands, 'from May 1st and hence was also called "Meifod."' It was of much slighter construction than the winter house or 'hendref,' that is, old house.

The Welsh 'fforch' was the equivalent of the Latin 'furca' and T. Hudson Turner, the historian of early English domestic architecture, considered that the five 'furcis' of which an armoury was to be made for King Henry III at his manor of Ludgershall, in the year 1246, were wooden piers[1].

The upright posts of gallows were also called 'furcae' and in an undated agreement between the Bishop of Rochester and the Abbot of St Edmunds, it was agreed that one fork ('lignum furcarum') should stand on the Bishop's land and the other on that of the Abbot[2]. So in the ballad of *Adam Bell*, the gallows which are prepared for William of Cloudesley are spoken of as 'a paire of galowes,' and in the nearly contemporary *Catholicon Anglicum*, 'a galowe' is translated in the singular, 'furca.' In the *Select Pleas of the Crown*[3] is a thirteenth-century illustration of the hanging of a man, in which the culprit is suspended from a cross-beam, of which the ends rest in the forked tops of two upright posts, and other illustrations are given in Wright's *History of Domestic Manners and Sentiments*. These upright posts are the 'furcae' and 'gallows' of the documents.

Mediæval illustrations show that cooking was done on a similar framework, and to-day the charcoal-burner in the South of England hangs his metal cooking-pot over a fire of sticks from a cross-piece resting in the forks of two uprights. Such posts are exactly the same as those of the Welsh summer-bothy, and prove that, at a time not very remote, houses were constructed in the same way as frames for cooking dinners and as gallows for hanging men. The construction of buildings had not become specialised and different from that of other cultural objects.

[1] *Domestic Architecture in England from the Conquest to the end of the Thirteenth Century*, p. 210.

[2] Thorpe's *Registrum Roffense*, p. 366.

[3] Edited for the Selden Society by F. W. Maitland.

Early rectangular houses were formed of one simple apartment; they were, in the old phrase, buildings 'of one bay.' Houses were anciently described by the number of bays which they contained, and a building of one bay was the poorest and least important. The Elizabethan bishop Hall describes a cottage in one of his *Satires* and exclaims:—'Of one bay's breadth. God wot! A silly cote[1].' When their ridges were carried on posts, they could be easily lengthened and so they had a distinct advantage over their circular or oval predecessors. As the writer has shown elsewhere[2], types of buildings, which have long been obsolete in England, still linger on in the Teutonic lands of the Continent. The position of England in the world was entirely changed by the discovery of America and the Cape route to the East: this was followed by an almost bloodless reformation, an insular freedom from foreign invasion, and—during the Commonwealth—a broadening of the basis of Government which, though premature and afterwards checked, was never altogether destroyed. These, in their turn, were succeeded in the eighteenth century by the enclosure of the common lands and the resulting adoption of individualism in agriculture, and for these reasons the culture stage of England rose much above that of Northern and Central Europe, which did not share in such advantages.

Houses with an axial row of upright posts between gable and gable were considered by Mejborg, the author of *Gamle Danske Hjem*, to be the simplest type of Danish buildings, and they are still to be found in Jutland and the Fünen group of islands. Instead of the early kind of posts with naturally forked ends, a more artificial fork was cut out of the top of the post, and in it the ridge-piece was fitted. As the practice of building developed, buildings were made more permanent in construction, and larger and heavier timbers were used. Some of the posts in the old Danish houses were of such a large size that a man could not girth them with his arms.

In Sweden also, in Western Gothland, such buildings are to be found, and one is described and illustrated in *Fata Buren*, a publication of the Skansen Museum. They are there known as 'mesula,' that is mid-column, constructions.

E. Gladbach showed that this early method of construction is

---

[1] Bk v, S. 1.
[2] *Journal of the Royal Institute of British Architects*, May 27th, 1911.

to be found also in South Germany and Switzerland, and the ridge-supporting post is variously named 'firstsaule' (ridge column), 'hochsaule' (high column) and 'hochstod' (high post)[1]. There is documentary evidence for their high antiquity in South Germany in a law which states the fines to be paid for the burning of houses[2]. The law speaks of 'that column by which the ridge-tree is supported, which they call "first sul."' It is glossed by Notker as 'magansul,' that is, strong column.

The poem *Beowulf* is the oldest in the English language, and in it the hero is said on one occasion to have stood on the 'stapol.' Moritz Heyne[3] considered that it was a central post, but its meaning is uncertain. R. Henning, in his book on the German house, says that the word is glossed 'patronus,' 'an appellation which calls to mind the king post ("könig der säulen") in a remarkable manner[4].'

Herr K. Rhamm[5], following Bruun, has described the remains of a temple in Iceland in which there was found a row of stones, fixed at intervals down the centre line of the building. They were very naturally considered to be good evidence for the former existence on the island of buildings with central posts. The posts may have been an archaism when the building was erected, because innovations are not readily adopted in sacred buildings.

T. Hudson Turner, whose knowledge of old English buildings was largely derived from contemporary records, says of the columns in the hall, that 'sometimes there appears to have been only one range of such supports which, extending longitudinally through the room, reached to and carried the ridge or crest of the roof[6].' The single row of posts or columns survived in halls divided by a single row of stone columns, even where there was no ridge-tree to be supported, as may be seen in a hall in the basement of the Romanesque keep of the castle of the counts at Ghent.

Nearly a century and a half ago, an antiquary of the period, Salusbury Brereton, travelling in Shropshire, visited the very ancient mansion of the Gatacres at Gatacre, and he found it so curious in its construction that he was moved to send the following description to the Society of Antiquaries, who published it in

---

[1] *Schweizer Holzstyl.*   [2] *Lex Bajuvariorum*, Liber I, cap. X, 6, 7.
[3] *Beowulf*, p. 95.   [4] *Deutsche Haus*, p. 172.   [5] *Urzeitliche Bauernhofe*, p. 578.
[6] *Domestic Architecture in England from the Conquest to the end of the Thirteenth Century*, p. 4.

Archæologia[1]. He says 'The hall was nearly an exact square, and truly remarkably constructed. At each corner and in the middle of each side, and in the center, was an immense oak tree, hewed nearly square, and without branches, set with their heads on large stones, laid about a foot deep in the ground, and with their roots uppermost, which roots, with a few rafters, formed a compleat arched roof. The floor was of oak boards, 3 in. thick, not sawed, but plainly chipped....The whole, I hear, is entirely pulled down since I saw it.' This is an example of the destruction which is always affecting our old buildings, and is every year making it more difficult to write the history of their development. There can be no doubt that such a building was a direct descendant of the old Germanic hall, which also had corner-posts, side-posts, and a central post, taller and more prominent than the others, and therefore very naturally called the king-post.

Such a building was also a prototype of the Gothic stone-vaulted apartment with a central pillar, of which one of the western side-chapels at Lincoln cathedral is an example. The central column and the vaults in stonework were merely a translation of the wooden king-post and boards used by the Northern peoples before the introduction of Latin civilisation, which taught them how to work in ashlar.

The period when such buildings, with their long ridges supported by central 'king-posts,' or by a row of posts of equal height, were erected in this country is so remote that they are only likely to survive, if at all, in buildings of the most humble kind. Enquiries which the writer has made among old people in North Wales have failed to show any knowledge of the term 'nennforch,' although the term 'nenpren' is still in use. The ridge-piece is still used and so its name remains, but the posts which supported it have gone, and their name is consequently forgotten. In a village, which the writer visited some years ago, on the west coast of Lancashire, there was a cottage which had had its ridge-tree carried by a post in the centre of the living-room, until recently, and it was considered to have been the most old-fashioned house in an old-fashioned village. The roof had been reconstructed, and as the post was inconvenient, it had been removed.

At Knaresborough manor-house an upright tree, which is carried

[1] III, p. 112.

through the ground-floor storey and cut off in a bedroom to form a small table, is said to have formerly run up to carry the roof.

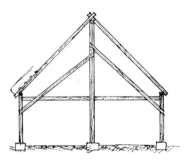

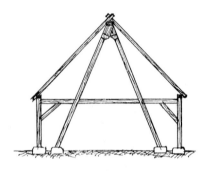

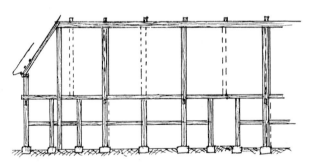

Fig. 4. Two Cross Sections (above) and Long Section of a barn at Herning, Jutland, Denmark. The ridge is supported by upright posts, and pairs of inclined posts, alternately.

A row of posts down the centre of a building was not only inconvenient in use, but was also weak in construction, and as a security against storms in certain buildings in Jutland, as in the villages of Salling and Herning, the single high posts are replaced alternatively by pairs of posts, with their feet apart, but meeting at the ridge: from this they are known in Danish as 'Straeksuler,' or 'Stridsuler'—that is, stretching or striding siles or posts. They are said to be very superior to the single-, or high-post buildings in resisting storms. The single upright columns are known as 'Skraasuler'—that is, upright siles or columns.

Mr Bernhard Olsen describes their construction as follows:—'The perpendicular posts ("skraasuler") are placed in a row alternately with the raking posts ("stridsuler"), then the ridge-beam is placed in position on them; after this the horizontal tie-beam is fixed, and the rafters are laid, and finally the studs ("stolper") are placed to form the wall.' It is important to notice that the walls are the last part of the building to be constructed. The illustrations (Fig. 4, opposite) show two complete cross-sections and a portion of a long section of a barn at Herning, in middle Jutland, with the 'skraasuler' alternating with the 'strid-suler.' Although this arrangement is an improvement in construction, it is no improvement in convenience, for the 'stridsuler,' alternating with the 'skraasuler,' must form a kind of gigantic *cheveux de frise* on the floor of the barn. The barn at Herning was built in the year 1802, and other buildings constructed in this primitive manner are of an even later date. The writer is not aware of the use of this form of construction in England, or elsewhere than in Jutland, but it is interesting as a stage in the development of construction, in the progress to higher forms.

A lean-to of boards which the writer has seen in a photograph of a scene in Dalmatia showed a connection between the old lean-to, earliest of all attempts at building, and the forked posts. The lean-to was formed by boards, of which the lower ends rested on the ground and the upper ends were supported by a 'ridge-pole.' The ends of the pole were, in their turn, dropped in to the forked ends of upright posts. The rectangular hut, such as the Welsh 'hafod,' may be considered as an obvious development of such a lean-to.

In North Wales the writer has found a tradition that the Welsh word 'ty,' meaning house, gave the name to the letter 'T,' because the early form of the letter was like a house with sloping sides and a central post or fork running up to the ridge.

Houses are often represented on tombs or gravestones, and the memorial stone or the grave itself may be given the same form as the house in which the dead man lived, in order that he may feel at home after death.   The so-called 'hog-back' stones, which are found in churchyards, or used as building materials in the mediæval churches of the North of England, are small copies in stone of the houses of their makers.   They are probably of the tenth century and upright posts are occasionally shown in the gable ends.

The upright posts were also called 'crotches' in this country and a translator, G. Sandys, who 'Englished' Ovid's *Metamorphoses* in the year 1640, rendered the passage 'furcae subierunt columnas' by 'To columns crotches grew.'   The word 'crotch,' with the meaning of upright post with a forked end, was carried by the English settlers to America.

CHAPTER IV

CURVED TREE PRINCIPALS

The German roof-hut—Its allies in Scandinavia and in England—The old cottages of Snowdonia—Principals formed of curved trees in England—Their names in the English dialects—Their use in buildings and their principal varieties—Preparation of timber where felled—Studs and stooths—Siles and gavelforks.

The Jutish method of supporting the ridge-tree by a pair of posts inclined to each other, alternately with single upright posts, must have been even more inconvenient than the older form of a single row of upright posts down the centre of the building, and its *raison d'être* lay in its superior strength. For experience has shown that gales which blew down the buildings with ' skraasuler ' left those with ' stridsuler ' standing. If the upright posts are discarded, and the ridge is supported by pairs of inclined posts only, so placed as to follow the slope of the roof surface, then the building is not only stronger in construction but its floor is freed from supports. The use of such a method of construction seems perfectly natural when once the use of rafters inclined against a ridge-pole had been adopted, and buildings formed in this manner have long been in use both in England and on the Continent. They are not of importance as examples of the art of architecture, but as survivals of a stage in the development of building construction.

Such buildings are still used for sheep shelters on the German heath-lands, from the Elbe to the moors on the Dutch frontier, and from their use they are known in Low German as 'Schapkoven.' As their roofs reach to the ground they are also known, from their shape, as ' Dach-hutten,' literally roof huts.

In Scandinavia they are called ' skali,' a word which is also used as we have seen for other forms of primitive buildings. In English the word became ' scale,' ' skell ' and ' skill,' and was applied in

Cumberland to temporary hovels for the shelter of shepherds, and possibly also for their sheep, as in Germany.

An illustration of a Scandinavian example (Fig. 5) from Jemtland is reproduced here ; the slanting poles which form the rafters are crossed at their summits, and so form a fork in which the ridge-pole is fitted, an idea obtained from the forked ends of the upright posts[1]. The illustration (Fig. 6) shows a Belgian example which the writer found and photographed at Huttegem, near Audenarde. It was used as a little store for the farm, and

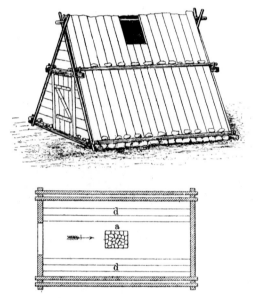

Fig. 5. A Scandinavian 'skali' from Jemtland. The roof
rests directly on the ground without side walls.

when the photograph was taken, it had just been put up by the farmer sitting in the front. He had formed it of light sapling boughs and covered it with tiles and thatch. The roof surface is curved and a masons' lodge, similarly constructed, but of larger size, is shown on a carving on a house at Middelburg, Holland, dated 1590[2]. Perhaps it was to a similar building that Wm Horman referred, when he wrote in the year 1519, 'His house hath a ridge like a shypp's bottom[3].'

---

[1] Reproduced by permission from Herr K. Rhamm's *Urzeitliche Bauernhofe*.
[2] Sydney R. Jones, *Old Houses in Holland*, p. 116.   [3] *Vulgaria*, p. 244.

A similar kind of building, if that is not too dignified a name, is constructed by charcoal burners for their temporary huts in the South of England and an example has been illustrated by Mr Edward Step in his *Sketches of Country Life*. The height and length are alike eight feet.   In this southern charcoal-burners' hut, of a type which is advanced in comparison with those of the North of England, we meet with the first example of that decrease in

Fig. 6. Farm store at Huttegem, near Audenarde, Belgium. This kind of building is a survival of a very ancient type of house. The roof rests directly on the ground.

culture, northward and westward, which is found in these islands for century after century.   In timber construction, the occurrence of this inferiority in the North and West has been ascribed to the immigrations of Flemings into the Eastern counties.   It is not however in architecture and building only that this peculiarity is found : it holds good also for articles of culture which have no connection with building.   Such are the hand-querns for grinding corn, which became obsolete in the North of England a century

ago, and much earlier in the South, but which are still in use in the
Highlands of Scotland and the West of Ireland.   Planets which
are at a greater distance from the sun receive less light than those
which are nearer to it, and through long ages, until modern times,
the Continent of Europe played the part of the sun for the culture
of these islands.   If immigrant Flemings improved timber building
in the Eastern Counties, and this has been by no means proved,
such an improvement is merely evidence of the influence of that
Continental culture which was ever less strongly felt as the
distance from the East and the South increased.

Fig. 7.   House at Scrivelsby, Lincolnshire.   The roof rests directly on the ground
and the ridge-tree is carried by pairs of inclined straight principals or crucks,
two in each gable.

At Scrivelsby, near Horncastle, in Lincolnshire, is an example
of a permanent house of the same form as the ' Schap-koven ' and
the charcoal-burners' hut of the South of England.   It is popularly
known as ' Teapot Hall ' (Fig. 7)[1].   Its timbers are stout and
carefully put together : at each angle of the building there is a
pair of sloping posts, which are connected by horizontal tie-beams,
just as are the sloping posts or rafters of the ' skali ' from Jemtland

---

[1] The illustration is reproduced, with the publishers' permission, from Mr S. O. Addy's
Evolution of the English House, in which the house is described.

before mentioned. Such a form of construction may be due to the Teutonic invaders of this country, and as it is used in both Germany and Scandinavia, it may have been introduced both by Saxons and Danes. It is not easy to assign a date to such a building as 'Teapot Hall,' but it can hardly be later than the end of the Middle Ages. It belongs to a time when the conception of the cottage as a permanent house was prevailing over the older idea of the removable hut. In the simplicity of its construction the building belongs to the old order ; in the evident desire of its builders for permanency, it belongs to the new.

'Teapot Hall' measures 19 feet in each dimension—length, breadth and height—and the pitch of the roof is steep in consequence. The use of the inconvenient upright king-posts had passed away, and to that extent the construction of 'Teapot Hall' is an improvement, but the form is still inconvenient from the rising of the roof directly from the ground, and a curious mode of construction, in which this difficulty was overcome, was in widespread use in this island from the Middle Ages and possibly before.

England was anciently possessed of vast forests of splendid oak-trees, of such a variety of growth, that the builder with the help of the forester could generally find such timber as he required. And so it was possible to find an oak-tree naturally bent in such a form that, when set up in the building, part would be upright and parallel to the wall and part sloping in line with the roof, or to find one with a great branch bent at such an angle to the bole that when the tree was set up in the building the branch would be in line with the roof. If such a tree were split and the halves placed opposite to each other, with the ends of the bent trunk or branch crossing at their summits, they would form a fork, in which the ridge-tree would securely rest.

Messrs H. Hughes and H. L. North have described[1] a number of old houses which they have discovered in North Wales, in which the wall post and the principal are formed of a bent tree in one piece, and the ridge-tree is dropped in the fork formed by the crossing principals. The authors say that there is good ground for thinking that the old churches of North Wales were roofed in this manner before the coming of the Anglo-Latin monks in the twelfth century, and the introduction by them of couple-close

[1] *Old Cottages of Snowdonia*, pp. 5 et seq.

roofs of English type. It is said that the nave of Llanfairfechan old church was roofed in this manner, and it must be remembered that the civil buildings were more antique in style than the religious of a similar costliness, in the Middle Ages, not only in England and Wales, but in France also, as Monsieur C. Enlart has shown[1].

However by personal enquiries among old men in North Wales, including old carpenters in Carnarvonshire, the writer has found a widespread tradition that these buildings were forced upon the Welsh by their English conquerors, and it was a friend of the Welsh—in one version the Black Prince, in another Henry Tudor —who removed the restraint and allowed them to build houses with walls. The conquerors imposed the use of such buildings because they could not be fortified, and Welsh ingenuity, by the use of the bent posts, kept the letter and broke the spirit of the law. The tradition, like other traditions in Celtic countries, is not very tangible, and probably the element of truth is an English superiority in building, which was followed by the Welsh, and an earlier use in England of upright walls in ordinary buildings. Usually the tree is fixed on the ground, but at Beudy'r Efail, behind Dinas Emrys, in the Gwynant Vale, the tree is fixed in the wall at some distance from the floor[2].

Messrs Hughes and North assign these cottages of Snowdonia to the fourteenth or early fifteenth century. This seems to be arrived at by comparison with dated buildings of a more elaborate kind, rather than from documents : they were obviously constructed at a period when small buildings were permanent, and sufficiently early for angularly bent oak-trees to be readily procured. An example is shown in Fig. 29.

Ty Gwyn, the house of Howel the Good in Carmarthenshire, where the Welsh laws were codified, is said, but without adequate proof, to have been formed as follows : Two rows of trees formed two rows of columns in its length, and each tree had a strong, sloping branch left uncut ; these uncut branches met and crossed in the middle, in this way a row of forks was made in which was laid a straight slender tree called the roof-tree. Low walls of wood and wattle were built in a line with the columns and at some distance away.

---

[1] *Manuel d'Archéologie Française*, II, p. 102.

[2] H. Hughes and H. L. North, *Old Cottages of Snowdonia*, pp. 19 and 20.

Mr F. Seebohm considered that the columns and principal rafters of the Welsh tribal house were formed in the same way as at Ty Gwyn and the Snowdonian cottages, that is by trees with the branches inclined to the pitch of the roof, but the evidence cannot be considered very satisfactory[1]. Mr John Ward has described and illustrated the remains of Romano-British 'basilical houses,' which are large apartments containing the foundations of two rows of columns of timber[2]. There is always a double

Fig. 8. Timber frame of an old building at Erdington, near Birmingham. The left portion is of 'post and truss' construction, but the principals of the portion to the right are bent trees in one piece.

and never a single row. Such apartments were probably derived from the 'tribal house,' and they may contain the origin of the constructions of the old cottages of Snowdonia, and of Ty Gwyn (if the latter were constructed in the manner claimed), as they seem to have been of native, rather than of Roman origin.

But the Welsh tradition tells that the bent-tree principals were introduced by the English, and the only specimens remaining in a church of which the writer knows are two pairs in the church porch at Conway, which is a building much more English than Welsh.

[1] *English Village Community*, p. 239.
[2] *Romano-British Buildings and Earthworks*, pp. 174 et seq.

In the Close Rolls of the reign of Henry III there is an item
'Six bent posts, one hundred and twenty rafters of a length of
twenty feet and six wall-plates[1].' The bent posts were possibly
like those found in the old cottages of Snowdonia. In the
gable end of a cottage at Weobley in Herefordshire there is a
cottage with bent-tree principals in the gable end, like those
of Conway church porch, but of a more rugged appearance[2].
Mr D. J. Roberts, of Erdington, Birmingham, has kindly supplied

Fig. 9.   End view of the bent tree principals of the old
building at Erdington shown in Fig. 8.

the writer with particulars and photographs (Figs. 8 and 9) of an
old building at Erdington, of which one part is constructed of
ordinary 'post and truss' construction (see chapter VI), but the
other part is of large trees angularly bent at the height of the wall-
plate, as in the North Wales examples. In the latter the ridge-tree

[1] 'vi postes tortos, cxx cheuerones de longitudine xx pedum et vi paunas' quoted in
the *Select Pleas of the Forest*, published by the Selden Society from *Rot. Litt. Claus*, II,
p. 106 *b*.
[2] Illustrated by J. Parkinson and E. A. Ould in *Old Cottages, Farm-Houses and
other half-timber buildings in Shropshire, Herefordshire, and Cheshire*. The authors give
no date and merely say that ' the huge timbers forming the gable...tell of the days when
oak was plentiful,' p. 25.

is supported in a fork formed by the crossing of the principals, but at Erdington, in one of the pairs of trees, the ridge is carried by a short collar, as in the example from Wigtwizzle, South Yorkshire, which is illustrated in Fig. 24 ; and in the others the collar is lower and rather longer, and the ridge-tree is carried by a very short post standing on a collar, somewhat as in the example from Falthwaite, also illustrated (see p. 57 and Fig. 27), but at Erdington the post is vertical, instead of inclined. The building at Erdington is im-

Fig. 10. Old building at Erdington, Birmingham, of which the framing is illustrated in Figs. 8 and 9, showing how an apparently uninteresting building may be of very interesting construction.

portant because it shows that angularly bent posts were used in the Midlands, and permits of a comparison with those of Snowdonia. Fig. 10 shows the building as it was before the pulling-down was begun. Yorkshire examples are unknown to the writer, but Moxon, himself a Yorkshireman, in his *Mechanick Exercises,* says that a knee is a piece of timber used in house carpentry, ' so cut that the Trunk and the Branch make an angle.'

The builders believed that the ridge-tree bore the weight of the roof, and its sufficient support was their solicitude. The method of supporting the ridge by the bent-tree principals was a conspicuous improvement as compared with the central post, but extensive as were the English oak-woods, the supply was limited of those trees which had a great branch, or had the trunk angularly bent, at the required angle and height, so it came about that the builders made use of curved or bowed trees. The pointed arch was fashionable in the Middle Ages—in fact, not merely fashionable, but universal—and so the builders chose trees of a shape such that

Fig. 11.   Destroyed building at Little Attercliffe, Sheffield, showing upright post carrying the end of tie-beam and the wall-plate.

they gave an approximation to a pointed arch when set up opposite to each other in the building to form principals. Trees curved in a perfect geometrical manner were rare, and so these old tree-principals are rarely as regular in their curves as the stone arches in the churches. As there was no straight and upright wall-piece in these bowed or arched principals, the builders had a difficulty in carrying the wall-plate. They solved the problem by framing an upright piece of timber to the back of the curved principal, at its foot as in Fig. 11 which shows the upright in a building at Little

Attercliffe, Sheffield: it is now destroyed. This carried the wall-plate and the end of the tie-beam in addition to forming an upright or 'quarter' in the wall itself.

Only the cathedrals, the castle keeps, the churches, and buildings of the greatest importance, were usually of stone in the Middle Ages: the ordinary buildings seem to have been constructed with the curved tree principals, and great numbers still remain. Examples are known to the writer in about half of the counties of England.

The form with the timber angularly bent is apparently older than the forms in which it is bowed or irregularly curved. This is to be inferred from their rare occurrence in England, and their prevalence in Wales, and the difficulty which there must have been in obtaining timber of the required shape. At the present time the great majority of the survivals are either bowed or irregularly curved.

According to Whitaker, the historian of much of the North of England, Crakehou, a village in Craven, ' principally belonged to Bolton Abbey, and was granted to the first Earl of Cumberland. From a survey made in the time of the first Earl, it appears that every house and barn stood upon crocks and was covered with thatch[1].' The 'crocks' are the curved tree principals, and the survey was apparently made about the middle of Queen Elizabeth's reign.

Whitaker, in the year 1818, was probably the first writer to describe this form of construction[2]. He wrote : ' It is difficult to assign with exactness the era of buildings which have no inscribed dates, and of whose erection there are no records. But perhaps we may refer the oldest specimens of architecture in wood now remaining among us to the time of Edward I. Instances of this style are found alike in the halls of some ancient manor-houses and their gigantic barns, which are little more rude than the other. The peculiar marks by which they are distinguished are these: the whole structure has been originally a frame of woodwork, independent of walls, the principals consisting of deep flat beams of massy oak, naturally curved, and of which each pair seems to have been sawed out of the same trunk. These spring from the ground and form a bold Gothic arch overhead ; the spars rest

[1] *History of Craven*, Third Edition, p. 528.
[2] *History of Whalley*, pp. 499 et seq.

upon a wall-plate, as that is again sustained by horizontal spurs, grooved into the principals. It was then of no importance that such erections consumed great quantities of the finest ship timber, and, indeed, the appearance of one of these rooms is precisely that of the hull of a great ship inverted and seen from within. Specimens of this most ancient style in perfection are the old hall of the manor-house at Samlesbury and the Lawsing Stedes barn at Whalley.' Whitaker says, ' By crooks are meant arched timbers ascending from the ground elsewhere to the roof. I much doubt whether there are any specimens of crooks in houses or barns later than the time of Henry VIII.'

The Rev. Dr J. Collingwood Bruce in his book on the Roman wall, in the year 1851, says that in the village of Walton, 'One of its dwellings furnishes a good specimen of the mode of cottage-building formerly prevalent in the North. The rafters of the house, which consist of large and rudely-shaped pieces of timber, instead of resting upon the walls, come down to the ground : they are tied together near the top by a transverse beam, and the mud walls, as the thatched roof, partially depend upon them for support[1].'

In Cumberland these arched principals or rafters reaching to the ground were called ' siles.' ' The lower ends were placed upon a dwarf wall, and, being of curved oak, the upper ends met at the ridge, and when erected they resembled a pair of whale's jaws[2].' According to Mr W. G. Collingwood, ' The oldest type of house in High Furness' (after the charcoal-burners' round hut) ' was that constructed of pairs of " siles "—great curved oaken beams set on the ground or on a low wall of stone[3].' Mr Collingwood says, ' The fashion of building stone halls for the greater families must have set the example to the yeomen, who, during the seventeenth century, turned their sile-framed cottages into byres and barns and built themselves solid stone dwellings[4].'

In Lancashire, south of Morecambe Bay, this method of construction was also usual; there ' The clam-stave-and-daub cottages were built with four or six whole trees as their framework ; these trees, or 'crooks,' were set in the ground and inclined towards each other, meeting at the ridge and tied together with cross-beams.

---

[1] p. 287.
[2] *English Dialect Dictionary, s.v.* Sile.
[3] ' High Furness' in *Memorials of Old Lancashire,* II, p. 179.
[4] Ibid. II, p. 180

Upon these the walls and gables were framed and wattled, and then filled in with clay and reeds and whitewashed[1].'

We have seen that this form of construction occurs in North Wales and in Northumberland, Cumberland, Yorkshire, Lancashire, and Lincolnshire in England. It also occurs in the counties of Westmorland, Durham, Chester, Derby, Stafford, Leicester, Northampton, Warwick, Worcester, Hereford, Gloucester, Oxford, and in Shropshire and Berkshire. This method of construction is also

Fig. 12. Map of England showing the distribution of buildings 'on crucks,' by closely-hatched lines. The counties of England where the existence of such buildings is unknown to the writer are shown by widely-hatched lines.

found in Scotland, even in the western islands; but Dr Joyce informed the writer that he did not know of it in Ireland, and the writer has been unable to find it in South-West Wales, which is also somewhat Irish in its associations.

The map of England (Fig. 12) shows by closely-hatched lines those counties in which examples of this curious and ancient method of building are known to the writer. It will be seen that

[1] Mr W. F. Price, 'Homes of the Yeomen and Peasantry' in *Memorials of Old Lancashire*, I, p. 255.

he has no evidence for its use East and South of a line drawn from the Wash to the Bristol Channel, with the exception of certain secluded hilly districts. This may be due either to Celtic influence, or, more probably, because the West and North were the most backward parts of the country in culture, and methods of construction were used there after they had been discarded by the builders of the South and East.

We have seen that in Cumberland and in High Furness the curved principals are called 'siles.' In Cumberland they are also named 'sile-blades' and 'sile-trees,' and the same terms are used in Scotland. 'Sile' is a widespread Teutonic word, meaning 'column,' and indicates that they were regarded as variants of the upright timber posts of the Germanic lands, just as the inclined posts were in Jutland. The use of the word in the North of England is as old as the fourteenth century, for the *Durham Halmote Rolls*[1] show that in the year 1364–5 one William Smith agreed to put up some buildings on a farm which he had taken at Nun-Stainton, and that the said William should have large timber in the wood of Aclay for six pairs of siles, ribs and firsts ('pro vi paribus de scilles, rybys, et firstis'). Earlier than this is an entry in the *Durham Account Rolls*[2] of 1338–40, which speaks of seven couples of siles 16 feet in length and four couples of siles 28 feet in length.

These curved timbers have other names in the North country dialects. In Northumberland and Durham they are known as levers, a word descended from Middle English. In the Finchale Priory Accounts there is a record of a payment, in the year 1481–2, for timber bought for a lever in the tenement of Robert Jackson[3].

It has been shown that the upright post with a forked end was called a fork, and in Durham, and in North Yorkshire, a pair of the ordinary bent tree principals is known as a 'pair of forks,' so that each timber is considered to be a fork, which gives some indication of the descent of the curved pair of posts from one upright post. The word 'fork' is found in English literature and in documents. Tusser, a rhyming agricultural writer of the sixteenth century, says, 'Let make an hous for beastis of forkis and boorde': and the Book (Liber) of the Abbot of Glastonbury of the year 1189

---

[1] p. 34. The *Halmote Rolls* have been published by the Surtees Society.
[2] II, p. 377. These *Rolls* have also been published by the Surtees Society.
[3] *Priory of Finchale*, p. ccclv. Also published by the Surtees Society.

mentions a barn at Wrington which has forks ('furcas') and a beam of the forks ('trabem furcarum'). The Close Rolls of the reign of Henry III contain an entry of two forks and two wall-plates ('ii furcas et ii paunas[1]'). Both the upright, straight posts and the slanting, curved crooks were 'forks' to the mediæval writers, and so it is impossible to say which kind was meant in the above quotations.

Farther to the South, the name 'forks' is used in Northamptonshire, and in that county many examples are said to still remain in ancient farmhouses and barns. They have also the name 'crucks' in Northamptonshire, and this is used in Westmorland and in the West Riding of Yorkshire; about Sheffield a building of this kind, with curved tree principals reaching to the floor, is said to be built 'on crucks.' In Lancashire and Cheshire · crooks,' a newer form of the word, is used, and in these two counties there are to be seen some of the largest and finest specimens in existence. A curved tree of the bowed shape used in ship-building is known in the trade as a 'crook,' and in the days when English warships were of wood, oak-trees of the same kind were advertised for sale as 'crooks' in the newspapers of the period.

The older form of the word is 'cruck,' and according to Lacomblet, the historian of the Lower Rhine, the ribs of boats on the river are known as 'crucks' ('curva ligna naviculae quae crucken dicuntur'). The use of the word is widespread in the Teutonic languages and its history is investigated in J. and W. Grimm's *Deutsches Wörterbuch*[2]. The *Wörterbuch* shows that the original meaning of the word was 'bend,' and that the present meaning of 'prop' is a later variation. At the present time, in 'Skaane,' as Mr Bernhard Olsen informs the writer, the word 'krykke' is only applied to the large post with a forked end, which carries the swingle-tree for the bucket of a well of the old-fashioned kind[3].

Our English word 'cruck' agrees with both the older and the later meanings, as a cruck is a curved prop.

The word 'cruck' has passed through the usual softening, like 'kirk' into 'church,' and has become 'crutch' and 'crotch,' the use of which, in building, is restricted to upright posts.

---

[1] *Rot. Litt. Claus*, I, p. 539 *b* quoted in *Select Pleas of the Forest*, published by the Selden Society.

[2] *s.v.* Krücke.

[3] An example is illustrated in *Billedbog fra Frilandsmuseet*.

Such timbers are also known as 'baulks,' but the same word is applied to all large timber in buildings.

In the *Agricultural Survey of Ayrshire* of the year 1811, a roof is said to be formed 'of strong cupples, termed "syles."' The writer by enquiry finds that the curved tree principals are called 'couplings' in Derbyshire, and that in North Wales they are called 'cwpl,' which is the same word in a Welsh dress. In Welsh the chevron of heraldry is also called 'cwpl.' Clare, Earl of Gloucester, is said to have borne three chevrons on his coat-of-arms, because he 'builded iii greate houses in one province[1]'; and in the Norman-French language, as used in England, 'cheverones' were ordinary rafters, which formed 'copules,' when fixed in a pair in the usual manner. In this, as in the Welsh tradition of the T, there is evidence for houses in England constructed after the manner of 'Teapot Hall' and the German 'Schap-koven.' King Henry III, in the thirty-fourth year of his reign, gave orders that the Queen's chamber, at his manor of Gillingham, Dorset, was to be lengthened by fifteen couples[2]: but these were of ordinary rafters, the 'close couples' of the writers on English Gothic architecture, and not pairs of forks or crucks.

In Levins' *Manipulus Vocabulorum* of the year 1570 the word 'bijuges' is translated as 'ye croks of a house'; but bijuges also meant double or twin yokes for a team. Dr C. M. Andrews says that before the Norman Conquest 'the ox-yoke was much like that in use at the present time, a semi-ellipse of bent wood, with ends joined by a bar, which was either tied or keyed[3].'

The district about Sheffield contains the remains of a great number of these buildings 'on crucks': this is due to its remoteness from main roads, and the life and movement which they bring, until comparatively recent times. Even now it is possible to travel out of Sheffield on an electric tram-car for a few miles and easily walk to farms where the flail not only hangs behind the barn-door, but is regularly used for thrashing. With one exception the many cruck buildings which the writer has visited in the neighbourhood of Sheffield are of one type, which only varies within narrow limits. But they are not only common in the

[1] Gerard Legh's *Accedens of Armory*, p. 180.

[2] T. Hudson Turner, *Domestic Architecture in England from the Conquest to the end of the Thirteenth Century*, p. 224.

[3] *Old English Manor*, p. 253.

country about Sheffield : they are to be found everywhere in the North of England and the Midlands, and in North Wales, especially where estates are small, or where landowners are poor, where the superior attractions of manufactures have been prejudicial to the improvement of agriculture, or where the district has been distant from the stir of affairs and the example of rebuilding.

Fig. 13. Building on crucks at Langsett, South Yorkshire, in process of demolition. The later walls and slates have been removed and only the timber framework remains. The outshot or aisle is probably an addition to the original construction.

If one of these buildings is entered for the purpose of ex-amination—for this a barn should be chosen, rather than a house, because in the latter the timber is usually hidden by flooring and wall-paper—we are at once impressed by the extremely archaic appearance of the interior, which is due to the great size of the timbers and the apparent roughness of the workmanship. The timber is often so irregular that it is surprising that the carpenter got his roof surfaces in true planes; and that he did so, and that work so rude should have lasted so long, is valuable evidence for the traditional skill of the English worker in wood.

Typical South Yorkshire buildings 'on crucks' are shown in Figs. 13—23. The crucks have their feet set usually on large, roughly-shaped stones (Figs. 13 and 14), or, very rarely, on

Fig. 14. Barn at Ewden, South Yorkshire. An addition has been made to the building in later times, but the original uprights carrying the ends of the tie-beams still remain.

pads of wood. The crucks are set up in pairs, and curve towards each other until they meet at their summits, where they support the ridge-tree. They have been happily compared to a pair of old-fashioned compasses, and also to the jaw-bones of a whale set up on end. Some little distance below the ridge, and on the back of the cruck, lies the 'side-tree' or purlin, of which there is usually one, rarely two, on each side of the roof. The purlin is usually held from slipping by a saddle joint on the back of the cruck, and it is propped by a pair of wind braces, which spring from a dove-tailed rebate on the back of the cruck at some distance below the purlin and are pegged down to it. In one case—at Bell Hagg, near Sheffield—the pair of wind braces is doubled, that is, the purlin is hung from above as well as propped from below. The wind braces are fixed into the backs of the crucks and the purlins by dovetailed ends fitted to the rebates: in order to avoid the meeting of the two wind braces on the back of the cruck, one wind brace is generally shorter than the other, but occasionally one wind brace is fixed on the top of the other. A tie-beam was framed

Fig. 15. Barn in the Little Don Valley, South Yorkshire. The illustration should be compared with Fig. 16 as it shows how a later casing may hide an earlier and rude internal framework.

across the pair of crucks at a height which was fixed by that of the side wall. As the pitch of the roof-surface was about 45° or rather less, it was necessary to fix the cruck at a steeper or flatter rake, according to the desire for higher or lower side walls. The height of the tie-beam varied accordingly, but this was rarely less than six feet and often much more. The beam projected at each end, and on it was placed the wall-plate, secured like the purlins, by a saddle joint.

At Somersall-Herbert, Derbyshire, there is a cottage which shows the transition from the non-projecting tie-beam of the type of 'Teapot Hall' to that just described, for at the level of the tie-beam 'a piece of oak has been attached by one wooden pin to the cruck, so as to continue the tie-beam, and projects horizontally as far as a vertical line taken upwards from the outside of the foot of the cruck[1]': the building has been considerably altered, and the walls are modern, as is now usual in buildings of the kind. Probably the 'horizontal spurs' described by Whitaker[2] were of the same intermediate type.

Fig. 16. The barn shown in Fig. 15 in process of demolition. The original tie-beams had been cut off, between the crucks, for internal convenience.

Each end of the tie-beam was propped underneath by a post, the foot of which was fixed to a kind of slot in the back of the cruck, a little distance above the stone base and pegged with the ordinary wooden pin (see Fig. 11).

In line with this post, at right angles to the pair of crucks, and at regular intervals—technically 'fixed at 18 inches centres'—were the studs which composed the framing of the original wall, and of which the upper ends were framed into the wall-plate. Very few of these upright pieces remain, for they have been replaced by stone,

---

[1] Letter from the Rector to the writer.   [2] *History of Whalley*, p. 500.

or even brick, walls in later times (Figs. 15 and 16), and this has occurred, not only about Sheffield, but in other districts also.

Canon J. C. Atkinson describes the buildings on crucks at Danby-in-Cleveland, where the crucks are known as 'forks[1].' They fail of being quite rectilineal throughout their length: four feet up is a curve, as in the timber of the stem of a boat, which allows a much steeper slope than if in one and the same straight line. From the curve 'the rafters are straight all the

Fig. 17. House 'Dike Side,' Midhope, South
Yorkshire (see also Figs. 18, 19 and 20).

way to the ridge-piece, and with the old tie-beam would form an almost exact equilateral triangle.' Dr Atkinson considered that it was 'obtrusively plain' that the existing side walls were an afterthought and that the buildings were of three periods; the crucks themselves were 'four, or maybe five, centuries old,' the later alterations to give more space dated from the middle of the seventeenth century, and the side walls were of a date intermediate between the crucks and the alterations. In South Yorkshire

---

[1] *Forty Years in a Moorland Parish*, p. 25.

enough remains to show that the carpenter worked on a rough rule for the scantling of the post as one-fourth that of the cruck, and the scantling of the studs was about one-half that of the posts. Sometimes when the tie-beam and wall-plate are high, there is evidence of an intermediate horizontal tie connecting the post and the cruck, in the form of a dovetail on the side of the cruck.

The feet of the crucks are often pierced by holes (Fig. 13) in the direction of the length of the building, and of greater height than width. At Dike Side, a house in the Little Don valley since destroyed for waterworks, the writer found that one hole was round

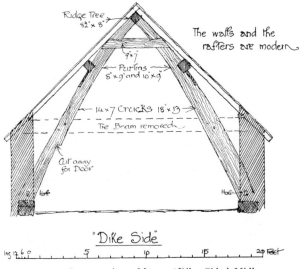

Fig. 18.   Cross section of house 'Dike Side,' Midhope,
South Yorkshire.

ended and another was rectangular (Figs. 17 to 19). The heights of the holes vary in the same building, which shows that they cannot have been mortices for rails, and as the pair of crucks was framed together on the ground, it is probable that they were used for passing ropes through, to keep the feet of the crucks together, and prevent them from 'opening' when they were being hoisted ; in addition they may have been used to insert wooden or iron crow-bars for raising the cruck feet on to the stone bases as they were of a suitable size for both purposes.

In a building in which there is a number of pairs of crucks, the tie-beam is invariably on the same side of all the pairs : if it is not, it is evidence that the crucks are not of the same date, or that they have been altered. If a pair of crucks is examined on the side across which the tie-beam runs, there are generally one or more parallel cuts to be found, on the surface of the tie-beam at the halving, and also on the cruck slightly above the tie-beam. On the first pair of crucks and their tie-beam there is one cut: on the second, two: and so on. It is thus possible to tell whether any pair of crucks is missing, and in one case, where a cottage had been

Fig. 19. End elevation of the crucks and other framework of the house "Dike Side," Midhope, South Yorkshire, in process of demolition.

built between buildings on crucks, an examination, by the writer, of the cuts on those remaining, showed that the crucks had once been continuous. It is also possible to tell from the figures whether the crucks are in their original position. From this careful marking of the separate pieces of timber it is evident that they were worked and fitted together away from the building site, and probably at the place where the wood was felled, in order to save the cost and trouble of the carriage of the portions which would be wasted in working up. A preliminary fitting together of building timber in the forest is widespread, and in Russia the building itself is put up roughly where the trees are felled, and then taken down and removed, log by log, to the intended site[1].

In the buildings about Sheffield all the pieces are pegged together with wooden pegs, and iron nails were not used in the framing, as they would have been useless in oak. Sometimes the pegs project so much that they may have been intended for perches for poultry or for owls. The pegs are either roughly polygonal or

Fig. 20. Side view of the house 'Dike Side,' Midhope, South Yorkshire. The later walls have been removed and the crucks and other framing remain. The site of the house is now occupied by a reservoir.

circular in section and driven in singly, not only in cruck buildings but in older English timber work generally, and the pair of flat opposed wedges used in old timber work on the Continent is only known to the writer in South Pembrokeshire, where there was a settlement of Flemings.

The pins were very strong and held the timber together in a manner which has defied the centuries. Palsgrave, who compiled a dictionary in the year 1530, expressed the feeling of the English carpenter when he wrote 'Je cheville. I shall pynne it so faste

[1] K. Rhamm, *Altslawische Wohnung*, p. 184.

with pynnes of yron and of wodde that it shall laste as longe as the tymber selfe.' The pins or pegs in the buildings *have* lasted 'as longe as the tymber selfe,' and are as firm as on the day when they were put in. When the building on crucks at Little Atter-cliffe, near Sheffield (Fig. 11), was to be pulled down, stout iron chains were fastened to the opposite feet of the pairs of crucks ·

Fig. 21. Barn at Cowley Manor, South Yorkshire, in process of demolition. The building is locally said to have been built, by a lady in the fourteenth century, to serve as a chapel.

then horses were attached to the chains, and pulled in opposite directions but the chains broke, before the wooden pins.

There is considerable variety in the shapes of the crucks in the Sheffield district, and it is unusual to find all the crucks exactly alike in the same building. As the country was heavily wooded, it is evident that an unfeeling regularity was not sought for by the

builders, and the effective survival of these buildings proves that
the mechanical accuracy aimed at by modern builders is not an
essential of successful construction. The crucks in the Sheffield
district may be classified according to their shape as more or less
arched or bowed (Fig. 21), regularly or irregularly wavy (Figs. 16
and 22) or reversed ogee, which approaches the shape of the
ogee-shaped stone arches which were fashionable in the fourteenth
century (Figs. 21 and 23). A pair of crucks in the English Lake

Fig. 22.   Interior of a barn at Dungworth, South Yorkshire,
with crucks of an irregularly wavy shape.

district has been illustrated[1], which are shaped like a true ogee,
that is, they curve outwards at the foot, and inwards towards
the ridge.

Between the tie-beam and the ridge there is often a collar, and
the absence or presence of this collar seems to be associated with
the method in which the ridge is fitted to the pair of crucks, which,
in its turn, is a guide to date.   (Figs. 22 and 23.)

From an examination of very many examples in the country
about Sheffield, the writer finds that there are three distinct

[1] H. S. Cowper, *History of Hawkshead.*

methods of fitting the tops of the crucks together to form a support for the ridge; there is, firstly, a type with the crucks roughly halved together, and projecting sufficiently for the ridge-piece to lie in the fork made by the crossing of their ends. The ridge-piece is fitted in this way in all the Welsh examples described by Messrs Hughes and North[1], and the writer believes that it is

Fig. 23.　Barn at Dwarriden, Ewden, South Yorkshire, with crucks of a reversed ogee shape, and a collar slightly above the level of the purlins.

the oldest form. A recently destroyed barn on crucks at the remote hamlet of Upper Midhope, South Yorkshire, had its ridge formed in this manner and it was the most archaic in appearance of any timber building which the writer has ever seen. In a variety of this a piece was cut out of the end of each cruck to form an artificial or partial fork, as at Little Attercliffe.

[1] *Old Cottages of Snowdonia*, p. 7 et seq.

The heads of the poles of the charcoal-burners' hut described in a former chapter merely leaned against each other without any fastening: in a joint of a more advanced type the posts were tied or bound together by some sort of cord or, as in this country, by withies. To tie joints together is a more primitive mode of fastening than to bore or to peg them, not only in building houses and barns, but in building boats also, and therefore a joint in carpentry which lends itself to being tied, may be considered as earlier in type than a joint which cannot be tied, and so is presumably earlier in date. Of the various methods of forming the ridges of the crucks, crossing and halving is that which best lends itself to tying. The oldest examples of the crucks in the Sheffield district are probably of the end of the fourteenth century, and ridges of this type are used with crucks of the reversed ogee shape. They are used also with crucks which have no collar between the ridge and the tie-beam, and in the Welsh examples the purlins do not appear to have wind braces.

The strength of timber was learned from practical experience, as was the strength of stonework, and the carpenter's art followed the same path from massive to light construction. A progressive economy in the use of materials is evident in the history of building, and the crucks of ogee form, crossed and halved at the ridge, are, on the whole, of larger scantling than other types : they are also more square in section, that is, they are made from a whole tree rather than from a tree split in half. (Fig. 16.)

Apparently it was found that the cross tie-beam at the level of the wall-plate was insufficient to hold the crucks together, and so, as has already been mentioned, a collar was used at or about the level of the purlins, of which there is usually one on each side.

It was also seen that economy might be effected by the omission of the crossing ends of the crucks, and by allowing them to touch only, thus forming an angle into which the ridge-piece was fitted. (Figs. 18 and 22.) This second form was not so strong at the ridge as the preceding one, and the intermediate tie was, in consequence, moved up gradually nearer and nearer to the ridge, until the collar became, in the end, merely a very short collar just under the ridge-piece and another, second collar, was placed at or about the level of the purlin. This short ridge-collar has often three mortice holes in it, and on the top of the tie-beam is a rough groove : this is evidence for the adoption in stud work of a usual method of fixing wattle stakes.

Further economy was possible with the timber, and in the third method the tops of the crucks were cut off horizontally, the short tie or collar was fixed across the top of them, and the ridge-tree laid flatly, instead of diagonally, as it had been in the previous methods.

It is possible to classify the crucks in the neighbourhood of Sheffield into types according to the relationship between ridge and collar as follows:

(i)   the crucks halved together at the top, with projecting ends forming a fork in which the ridge lies, but without a collar (Fig. 21):

(ii)   the same with the addition of a collar connecting the crucks at about the level of the purlins:

(iii)   the crucks merely touching at their tops, and forming an angle in which the ridge-tree rests, and with the collar in any position between the purlins and the ridge (Figs. 18 and 22): usually the higher the collar, the later is the date of the construction:

(iv)   as the last, but with an additional collar below the first:

(v)   the tops of the crucks cut off level, the short tie, or collar, mortised on them, and on it the ridge-tree fixed flatly instead of diagonally, as in the other types.

In November, 1912, some very old cottages at Little Attercliffe, Sheffield, were demolished. They were built on crucks, of which there were two pairs, contemporary in date, but not in type, as the junctions of the tops of the crucks differed. In one pair the junction was of the early crossed type: in the other, of the more developed type, with a short tie. Thus these crucks at Little Attercliffe were an example of the overlapping in methods which renders caution necessary in the dating of ordinary buildings and they showed how the craftsman compromised between the old method and the new. The same difference in types was to be seen in the crucks of the barn or 'chapel' at Cowley Manor (Fig. 21). The crucks were contemporary in this case also, as the numbering showed.

These crucks at Little Attercliffe, the only examples in the city of Sheffield which the writer has been able to examine with the ease which attends demolition, were also interesting, because they showed the variations in a recognised manner of building which arose from the personal inventiveness of the builder. Here he used cleets, pegged to the crucks, to support the purlins, instead of cutting away the backs of the crucks as much as was usual. He also used more than the ordinary freedom in the disposition of the wind braces, and he carefully numbered every part of the framing of the crucks and tie-beams on both sides, and the wind braces and purlins also. At Little Attercliffe the posts under the ends of the tie-beams remained and, as they have usually been destroyed in the examples in the district, they are illustrated in Fig. 11. The print also shows the dovetailed rebate on the back of the cruck, in which the wind brace was fixed. The rebate at the end of the tie-beam seemed to have been altered.

A comparison between the cruck buildings of South Yorkshire and the Jutish buildings with 'stridsuler' will show that they have no relation to each other. The principal differences in the Jutish buildings are the fixing of the ridge-tree by pieces of spiked timber to the ends of the slanting 'siles,' the straight 'siles' instead of the curved crucks, and the independent posts with the wall-plate carried on their tops, instead of the uprights fixed to the backs of the crucks, and the wall-plates resting on the tie-beams, as in the South Yorkshire examples.

The method of carrying the ridge beams by curved trees, apparently British in its inception, had the merit of freeing the building from the inconvenient central supports used in earlier times. It was natural that varieties should occur, and that it should influence more elaborate constructions during its long and general use.

Closely related animals or plants found in lands far apart are usually considered to be the survivors of a type once more widely distributed; but when an unusual kind of building or social custom is found in a remote district, it is difficult to decide whether it is a survival of what was once general and usual, and due to the remoteness of the district from outside influences, or whether it is merely a local development due to that same remoteness or to some special cause.

At Wigtwizzle, a remote hamlet on the edge of the South Yorkshire moorland, are the remains of a cruck building, intermediate in its form between the normal examples of the district and the angularly bent and apparently older forms. The crucks are now almost hidden in a farm building of later date, and only one pair can be measured (Fig. 24). The crucks are bent somewhat as in the Welsh examples but there is a curve instead of a sharp angle, where the roof slope joins the wall : the tie-beam does not project past the crucks, and it is not straight, but rises in the centre. There is a chamfered spandrel piece which is worked on the solid of the crucks but is formed by pieces let in, underneath the tie-beam. There are two purlins on each cruck, as in the Welsh

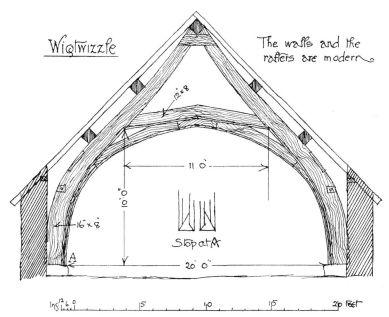

Fig. 24. Wigtwizzle, South Yorkshire. The framing is of an unusual kind, and is now almost hidden by a farm building of later date.

examples, instead of the usual single purlin of the district. The ridge is formed by a short tie at the top of the crucks, the ridge-tree is laid diagonally and there is no intermediate collar. The illustration shows that the original walls must have been very low in height.

At Luntley Court, in Herefordshire, in a barn, there is a some-what elaborate variety, shown in Fig. 25[1]. There is no lower tie, and the Gothic foliations show that it once belonged to a building

---

[1] *Transactions of the Woolhope Field Club*, 1900-2. The sketch by Mr Clarke of Hereford, has been redrawn and is reproduced here by Mr Clarke's permission.

of some importance, probably of the fourteenth century. From
the upper tie run two struts which prop the upper parts of the
crucks. The method here shown of supporting the crucks, or
principal rafters, from a tie lasted a long time in English building
construction, and it is an example of the English use of a tie as a
beam in both tension and compression.

Other types of crucks have been found in North Wales by
Messrs Hughes and North[1]. Thus at the Old Plas, Llanfairfechan,
the crucks are bowed rather than bent, and there is no tie-beam:

Fig. 25. Timber Framing at Luntley Court, Herefordshire. The
principals are crucks, decoratively treated.

but, instead, there are two collars which cross each other at an
obtuse angle, each running from the cruck, just above the wall-plate,
upwards to the opposite cruck, at the height of the upper purlin.
The crucks project from the walls into the interior of the building,
and their feet are placed on stones above the floor.

At Merchlyn Bach, near Ro Wen, in a building which unfortunately has been destroyed, the crucks or principals, and also the

---

[1] These types are fully described by the authors in the *Old Cottages of Snowdonia*,
p. 9 et seq.

tie, were curved. This gave the appearance of an arch, such as is given by the spandrel pieces at Wigtwizzle, and is often seen in early wooden doorways, where the side posts and the head, or lintel, are cut to a curve to give the appearance of an arch.

Another example at Tai Fry, near Hafod y Maidd, Cerrig y Druidion, had two vertical posts springing off the horizontal tie: it has shared the usual fate of interesting buildings and has been destroyed.

The absence of wind braces in the Snowdonian examples may have been due to the shortness or narrowness of the bays, as the object of the so-called wind brace is to decrease the bearing of the purlin and act as a strut. At Iscoed Isaf, at the head of the Nantlle valley, the purlins are slightly supported by little brackets from the principal. The tendency to variation which is shown by the Welsh examples seems to indicate that this method of construction was drawing to a close when they were built, and that the builders were seeking, somewhat blindly, for new developments.

Gudmundsson, in his book *Privatboligen paa Island i Sagatiden*[1], considered that the roof of the Old Norse hall was constructed with ridge-tree and purlins, and was the converse of the Low German 'sparrendach.' This roof, which he called 'aastag,' had a row of upright posts ('sul' or 'sula') under each purlin, and thereby formed a building with a central and two side aisles, instead of the more primitive single row of upright posts carrying the ridge-tree, and dividing the building into two equal aisles. From purlin to purlin, on the tops of the posts, ran a cross beam ('vagl' or 'vaglbiti'), and in the centre of this was a short post ('dvergr'= dwarf), which carried the small ridge piece. There were other rows of shorter posts ('utstafir') fixed in the wall. (Fig. 26.)

It has been suggested that we have the English descendants of the Old Norse halls in the old barns and other buildings divided longitudinally into three aisles, like a church, by two rows of wooden posts or columns. Some information as to this should be obtained from a comparison of the English dialect words with the names in other Teutonic languages. There is a widespread Teutonic word for column or post which assumes somewhat different forms in various languages: in Modern High German it is 'säule,' in Modern Danish 'sul,' in Old Norse 'sul' or 'sula.' Its English form is 'sile,' and

[1] pp. 116 to 125.

it occurs in fourteenth century documents with the meaning of
'cruck': the word is still used in the dialects of Scotland and the
northern counties of England,
but in none has it the meaning
of vertical column. It always
means either a cruck or a principal rafter. In Cornwall the form
of the word was 'sills,' and it
occurs with the meaning of principals in a memorandum, dated
1772, made by the then vicar of
Zennor[1]. One line of the ancestry of our modern principal
rafter was by development from
the cruck, and the restriction of
meaning of the word 'sile' in the
dialects implies that crucks and
principal rafters were regarded

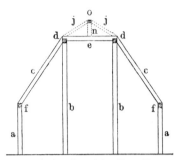

a a útstaf-r; b b súla, súl;
c c skorðrapt-r; d d langáss, hlíðass,
brúnáss; e vagl, vaglbiti; n dverg-r;
j rapt-r; o moenáss.

Fig. 26.   Section of an old Norse hall
according to Gudmundsson showing
timber framework.

as the equivalents of the upright wooden posts by which the roof
was supported in times older than any existing documents.

'Stud' or 'stooth' is another old Teutonic word for a wooden
post or column. Bede[2] gives an account of the death of St Aidan
as he leaned against a post of his church: the word post is
'destina' in the original and in the Anglo-Saxon translation of
the history it is rendered 'studu.' In Aelfric's glossary the words
'destina,' 'fulcimen' and 'postis' are all translated by 'stiper.'
In an Anglo-Saxon vocabulary of the tenth or the eleventh
century, the word 'postis' is translated by 'durstodl,' and in
Latin and Anglo-Saxon glosses of the eleventh century by
'durustod.' Both words mean doorposts in Modern English, and
'stud' and 'stooth' are now restricted to the upright posts in a
lath-and-plaster partition. 'To stuthe' is translated 'stipare' in
the fifteenth century *Catholicon Anglicum* and 'stiper' is now a
dialect name for the piece of wood placed upright in the doorway
of a barn and against which the double doors are shut.

It will thus be seen that the old names for upright posts, as far
as they are a guide, do not indicate that the buildings on posts are
descendants of either the Norse halls or of those English buildings

---

[1] J. T. Blight, *Churches of West Cornwall*, p. 153.
[2] *Ecclesiastical History of the English People*, Bk III, chap. XVII.

of former times which were the ancestors of the buildings on crucks. In a fine barn, near Falthwaite, South Yorkshire, the crucks at each end of the building are cut short at their tops, and connected by a cross-beam at some distance below the ridge (Fig. 27). This cross-beam carries a short and slight post, inclined inwards and supporting the ridge-tree, which does not reach as far as the end

Fig. 27. Barn at Falthwaite, South Yorkshire. The skilful manner in which the carpenters made use of unstraight timber in the tie beams, etc., should be noticed.

pairs of crucks. The other crucks are normal, and the arrangement of the end pairs of crucks may have been a repair after decay, or again, it may have been original and intended for a thatched and hipped end, as there are cottages near Gloucester with such thatched and hipped ends in which the crucks do not reach as high as the ridge, and are connected by a beam. Cottages with crucks of the normal type are also to be found in the county, and the illustration (Fig. 28) shows their appearance in the gable end. The arrangement at the ends of the Falthwaite barn, whatever may have been its purpose or its origin, is that of the Old Norse hall, with the dwarf post ('dvergr') and cross piece ('vagl'), as described by Gudmundsson[1], except that inclined crucks had superseded upright posts at the time of the construction of the barn.

Fig. 28. Cottages built on crucks which show on the gable end,
at Dymock, Gloucestershire.

It is possible that, when the work was carried out, some building of the Old Norse hall type still existed in the neighbourhood.

An arrangement of construction is illustrated in the *Memorials of Old Lancashire*, which is intermediate between that of the Old Norse hall, and that with true crucks meeting at the ridge. The book illustration shows a cottage at Farnworth, now destroyed, at each corner of which were two upright and straight corner posts,

[1] *Privatboligen paa Island i Sagatiden*, p. 122

and within these were the crucks, like a very flat cyma in shape and reaching only to the purlins. This rare variety gives a clue to the reversed ogee shape of the older crucks, for such crucks are merely like those of the Farnworth type prolonged to carry the ridge. These crucks at Farnworth are as much posts as crucks and are valuable as showing the intermediate form between posts and crucks. As there is no satisfactory origin known for the 'ogee' arches in stone in the mediæval buildings of stone, it is possible that they were copied from wooden examples.

A method of building construction is as little descended from only one ancestor as is a man or a horse, and the cruck building was the British descendant of the 'dach hutte' and the Norse hall. The Jutish 'stridsuler' have, as already mentioned, another line of descent, and the cruck buildings seem to be as peculiarly British as the bull dog or the red grouse. There are boat-shaped buildings in Europe supported on curved timbers, but there is no evidence to show that our cruck buildings ever were boat-shaped, and the type with angularly-bent crucks, so usual in North Wales, must be regarded as the older, for the fewer and simpler are the joints in carpentry construction the older is the type. The evidence for its earlier date is its occurrence in Wales, its rarity in England and the simplicity of its construction.

The apparent absence of any recognised Mediæval Latin name for crucks is another evidence of their insular origin. The earliest known use of the word 'cruck' occurs in a deed of the year 1432, relating to the West Riding of Yorkshire[1]. In it the crucks are called 'laquearibus contiguis' and the writer of the document thought it advisable to add, 'Anglice, viij crukkes.' Levins, as we have seen, translated crucks by 'bijuges.' 'Laquear' usually meant ridge-tree, but in the *Ortus Vocabulorum,* a work of the sixteenth century, 'laquear,' 'laqueare,' and 'laquearium' are explained as the joining of the beams at the top of a house ('conjunctio trabium in summitate domus[1]').

In a preceding chapter certain buildings in Jutland were described in which the ridge is carried by alternate upright and slanting wooden posts, and in the Middle Ages there were English buildings which possessed both upright crucks and posts. Mr S. O. Addy[2] has translated a portion of a lease of the

[1] H. Ling Roth, *Yorkshire Coiners,* p. 155.
[2] *Evolution of the English House,* p. 59.

year 1392, from the *Feodarium Prioratus Dunelm*[1], as follows :
' All the buildings of the said messuage—namely, a house, called
the fire house, containing five couples of siles and two gavelforks :
a storehouse for grain ("grangium "), containing three couples of
siles and two gavelforks : and another storehouse for grain, containing one couple of siles and two gavelforks : a little house to
the west of the said fire house, containing three couples of siles
and two gavelforks.'   Here the couples, or pairs, of siles vary
from one to five in number, but it will be observed that whatever is the number of the siles, there are never more and never
less than two gavelforks in each building.   The siles and the
gavelforks were the most important part of the construction, and
by them the building was described.

This combination of siles and gavelforks occurs in many other
documents of the fourteenth century, relating to the North of
England.   In all the buildings there are always two gavelforks,
whatever may be the number of the pairs of siles.   The dialects
show that the siles were crucks, but the meaning of the word
gavelforks is more doubtful.   In the fifteenth century *Catholicon
Anglicum* the ' gavelle of a howse ' is translated ' frontispicium,'
that is, a house front.   The contract for the rebuilding of Catterick
Church, Yorkshire, in the reign of King Henry V, states that
' the forseid Richard sall make a window in the gavel of fife
lightes accordant to the height of the kirke[2].'   In the year 1788
the word ' geeavle ' meant the gable of a building[3], and ' gavel ' is
said to be still applied to a gable in Scotland[4].   Evidently the
' gavel ' was the gable end and the ' gavelforks ' were upright
posts with a forked top set in the gable wall.   This is con-
firmed by the contract for building the bridge at Catterick in the
year 1421, in which it was agreed that the masons would ' make
a luge of tre of iiii. romes of siles and ii. hen forkes[5].'   'Hen' is
obviously the equivalent of the Welsh ' nen,' already discussed,
and the gavelforks and henforks are the same as the nenffyrch of
the Welsh triads and Laws.   The Welsh, in their turn, borrowed
the word ' gavel ' from the English.

[1] p. 167.   Published by the Surtees Society.
[2] T. D. Whitaker, *History of Richmondshire*, II, p. 25.
[3] W. Marshall, *Rural Economy of Yorkshire*, II, p. 322.
[4] *Dictionary of the Architectural Publication Society*, s.v. Gavel.
[5] Ibid. s.v. Henforkes.

It is evident that the fourteenth century, a period of cultural expansion, was the time of transition in the North of England from the method of supporting the ridge-tree by a single post to that of propping it by a pair of forks or crucks, and it is also evident that while the Danish 'stridsuler' were devised to strengthen the building, the English crucks were used for the purpose of convenience.

CHAPTER V

CURVED TREE PRINCIPALS (*Continued*)

Separation of the wall post and the principal rafter—Economy in the use of
materials—Mediæval roofs in churches and large barns—Ancient Irish
oratories of stone—The pointed arch.

The later crucks were of various shapes—bent, irregularly
waved, and so on—and this variety was probably due to the
increasing difficulty in obtaining trees of the shape required.   In

Fig. 29.   Old cottage near Conway, in Carnarvonshire,
in which the principals are formed of angularly bent
posts and rafters in one piece resting upon the
ground.

North Wales, where angularly bent crucks only were used, an
escape was made from this difficulty by forming the wall-piece and
the principal rafter of two separate pieces of timber, so jointed
at the angle that they appeared to be one.   The owner—and

occupier—of one of these buildings in North Wales told the writer that he did not think there were any buildings in existence in which the principal rafter and the upright were in the same piece of timber, although the writer had just examined an example not very far away, shown in Fig. 29, and the student of old ordinary buildings may here be warned against accepting statements without the evidence of a personal examination. Messrs Hughes and

Fig. 30. House near Conway, in Carnarvonshire. The principals are similar in appearance to those shown in Fig. 31, but the posts and rafters are separate, and connected by a triangular piece to strengthen the joint.

North have illustrated examples[1] which they consider to be of the late fifteenth or early sixteenth century. There is no cross tie-beam at the level of the wall-plate, but a collar about half way up the principal rafter. In the example illustrated in Fig. 30, Cymryd, as at Luntley Court, two raking struts cut into simple Gothic foliations spring from this collar to support the principal rafters, which cross at the ridge, forming a fork in which the ridge

[1] *Old Cottages of Snowdonia*, p. 16 et seq.

piece is supported. The wall piece does not extend down to the floor, and the joint between it and the principal rafter is strengthened by a triangular piece secured to each.

Similar buildings, in which the principal rafter has a wall piece at its foot which in some cases reaches quite to the floor and in others not, are common in England and are usually of the fifteenth century.

Fig. 31. Ruined cottage at Cae Crwn, Glanmorfa, Portmadoc, North Wales. The crucks, or principals, curved only at their feet, approximate in arrangement to a collar beam-truss.

It has been shown that some early timber buildings had their wooden supports fixed in holes in the ground, and according to Moritz Heyne[1] it was the custom on the Continent to char the portions in the ground, to make them less liable to decay through damp.

Some Anglo-Norman stone buildings are built without foundations, and if stone walls were thought suitable to be placed directly on the surface of the ground, then wooden posts might

[1] *Deutsche Wohnungswesen*, p. 18.

also be set up on the ground. In a house, possibly of the
fourteenth century, at Edmundbyers, Durham, the feet of the
crucks, or 'forcs' were either resting on the ground or sunk a few
inches below the surface[1], and at Danby-in-Cleveland their ends
were placed on the natural ground or on a large flat stone[2]. In
the Sheffield district it is evident that many of the stones on
which the crucks stand are later insertions, and in such cases the

Fig. 32. Barn at Southey, near Sheffield. The
crucks here have been developed into the
principal rafters of an ordinary roof-truss.

crucks had probably been placed on the ground originally, and
their feet had rotted in the course of time.

It was then merely a matter for observation by the builders to
see that it was an advantage to raise the feet of the crucks above
the ground, for this prevented decay, and it also gave a further
opportunity for that economy in the use of materials which has
been an underlying motive in the development of building construction as it has been in architectural design, for the higher the

[1] Rev. W. Featherstonehaugh in *Archæologia Aeliana*, vol. XXII, part ii.
[2] Rev. J. C. Atkinson, *Forty Years in a Moorland Parish*, pp. 23 et seq.

foot of the cruck was raised the shorter and cheaper would it be. A building in which the crucks are raised well above the ground is to be seen at Kimberworth Hill Top, near Rotherham, Yorkshire ; and in a ruined and secluded building which the writer found at Cae Crwn, Glanmorfa, near Portmadoc, the curve of the crucks is reduced to a bend at the foot of the principal rafter. It is shown in Fig. 31, but the situation did not admit of a satisfactory photograph. The English seem always to have regarded the tie-beam as capable of being used both as a tie and also as a weight carrying beam, and the economically minded builder of a barn at Southey, near Sheffield, in the early part of the seventeenth century conceived the happy idea of lifting the feet of the crucks on to the tie-beam, which, for the purpose, he made stronger than was usual (Fig. 32). The upper portion is straight from the purlin to the ridge, as was usual with crucks, and the lower portion is curved. The crucks have become simply the principal rafters of a roof-truss, and the curved foot is only a survival from the time when the curved principal stood on the ground. At the end of the Middle Ages, and before the application of mathematics to construction in the time of Wren, the English carpenter seems to have been feeling out for fresh methods, in a manner analogous to that of the mason several centuries earlier, in those experiments in design and detail which accompanied the Transition from Romanesque to Gothic.

Of these trials in new manners the roofs at Cymryd and Cae Crwn and Southey afford typical examples.

Unlike the English mason, the carpenter was little, if at all, affected by the French influences which came into building with the Norman Conquest and retained in a greater measure the English air for freedom in design and for individualism in construction.

At Crakehou, as we have seen, every house was built on crucks, and this method of construction was everywhere so usual that they must have formed an environment from which the carpenter could hardly have been able to detach himself. Their influence in the Middle Ages may still be seen here and there in the roofs of more important buildings, for the carpenters' work is less ordered and more individual than the work of the contemporary mediæval masons[1]. In the Middle Ages the craft of the mason was in

[1] Examples have been illustrated by Mr F. B. Andrews, in his book *Mediæval or Tithe Barns*.

advance of that of the carpenter. As an example, R. and J. A. Brandon[1] pointed out that the great Anglo-Norman roofs were constructed simply of trees with their branches cut off.

Morton Church in Lincolnshire had a roof of which the outline was evidently derived from the curved crucks of the period, with this difference, that in the church roof the crucks are halved by the tie-beam instead of being halved to it (Fig. 33), and the feet of the crucks are carried on a corbel built into the wall[2]. In this way the carpenter solved the problem of roofing a building which was higher than those to which he was accustomed. In the well-known barn at Bradford-on-Avon[3], generally considered to be

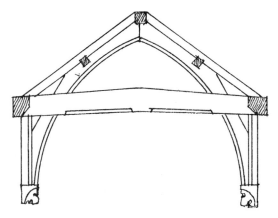

Fig. 33. Roof-truss in Morton Church, Lincolnshire, an example of the influence of cruck construction in the more important buildings.

of the first half of the fourteenth century, the principal rafters curve at their feet somewhat like the crucks at Wigtwizzle, illustrated in Fig. 24, but on a much larger scale. For this work of a large size the carpenter based his design on the ordinary buildings which he was accustomed to construct, and the feet of the principals have spread outwards in consequence.

In a farm building at Bradway, Derbyshire, the writer found pairs of crucks which had survived seventeenth century alterations. In them the carpenter had the problem of roofing side walls higher in proportion to the width of the building than was usual,

[1] *An Analysis of Gothick Architecture*, I, p. 91.

[2] The illustration has been redrawn from Messrs R. and J. A. Brandon's *Open Timber Roofs of the Middle Ages*.   [3] Illustrated in *Mediæval or Tithe Barns*.

and he had solved the difficulty in an unusual manner. A piece of wood, the upper part of which tapered to a feather edge, was pegged to the back of the cruck at some distance below the ridge. The usual purlins and wind braces were fixed into the back of this piece of wood, which was contemporary with the crucks and served as a principal rafter of a very primitive kind. This is shown in

Fig. 34. Old Farm Building at Bradway, Norton, North
Derbyshire, showing rudimentary principal rafter fixed
to the back of the cruck.

Fig. 34: the floor is that of an upper storey which was put in during the seventeenth century alterations. The original tie-beam is now inconvenient, and the crucks run down to the floor of the storey below.

Under the floor shown, and below the tie-beam, is a later heavy beam which carries the joists of the floor. There is a considerable amount of notation at Bradway. The purlin is numbered, and notched for the piece on the back of the cruck, and in another pair the cruck itself and the piece in the back are both numbered. The wind braces spring from the piece on the back of the cruck, and there are no rebates for them on the cruck itself.

In North Wales the writer has found other examples of this rudimentary primitive rafter. At Cae Crwn, already mentioned, the plane of the roof is continued to the wall by a piece of timber pegged to the back of the cruck at the curve (Fig. 31). In another example, Tyddyn Cynal, lying secluded where the green Welsh hills meet the wide Conway sands, the crucks in one of the pairs are not only angularly bent in the Welsh manner, but they are also of reversed ogee form and there is a piece of timber pegged on the top to bring them to a straight line, in curious endeavour to compound fashion with practical convenience[1].

The Herefordshire example from Luntley Court, illustrated in Fig. 25, is an advance upon that at Bradway, for the principal rafter reaches from ridge to wall-plate, but the cruck still remains the principal timber of the roof. Further progress in the importance of the principal rafter led to the 'arch braced' roof so common in the Middle Ages, in which the cruck is reduced to an arched brace, which does not extend to the ground.

At this place there may be mentioned a Welsh farm house of an extremely archaic appearance, situated between Strata Florida Abbey and Pontrhyddfendigaid, which has been described by Mr T. W. Williams and illustrated by Mr Worthington G. Smith in *Archæologia Cambrensis*[2]. The date of erection was supposed to be prior to the dissolution of the monasteries. The walls were of rough rubble, and the timber, including that of the doors, was unsquared. The cow, the pig and the poultry shared the house with the farmer and his family. The plan is shown in Fig. 35 and an interior view in Fig. 36. The post supporting the ridge in the centre is a survival of the 'nenfforch' (see p. 15) while the side posts are perhaps copied from English post and truss buildings. The pieces pegged underneath the cruck-like principal rafters are curious; that on the right runs through the wall into the forked top of a

---

[1] This is not shown in the pair of crucks in the illustration (Fig. 29).

[2] XVI (1899), pp. 320–325. The illustrations are reproduced here by permission of the Cambrian Archæological Association. The farmhouse had shared the usual fate of interesting buildings, when it was described in 1899.

free standing external post, as shown on the plan. The internal screens were of wattlework, 'probably plastered with mud.'

Some students of cruck buildings have seen, in their curved principals, evidence for a descent from the ships of our seafaring ancestors: others, again, observing the resemblance of a pair of curved crucks to a pointed arch, have seen in them the origin of Gothic architecture. Very few crucks are regularly curved like a Gothic arch of stone. They are either bent in the centre, or near to the base, or are shouldered about the height of the wall-plate, or they are shaped like an ogee, or are even of an irregularly wavy

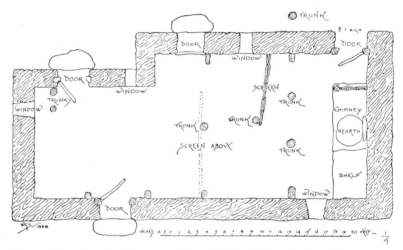

Fig. 35. Plan of a farm-house near Strata Florida Abbey, Cardiganshire, destroyed before 1899. The construction and arrangement are of a very primitive character. The 'trunk' standing free outside the wall carried the end of the right-hand principal rafter in Fig. 36.

form. It is more probable that the form of the curved or arched crucks was influenced by the regularly curved Gothic arches, and that the straight and elbowed crucks are the older forms. The crucks with a quick curve were evidently selected to afford more space at their lower part inside the building, and it is possible that the four-centred stone arch, a peculiarly English form, was derived from them.

A century ago, George Saunders, an architect of the time, clearly showed that the use of the pointed arch arose from the necessities of vaulting, and to-day that is the accepted theory

of its origin[1]. If the earliest pointed arches were derived from
wooden crucks, their position and use would at first be similar,
that is, we should expect to find them used as principals to carry
wooden roofs ; but the earliest pointed arches in England are used
in vaulting, and at Durham cathedral, where those early pointed
arches had their home, we find that the wooden roofs over the nave
aisles are carried by half round arches, and the earlier roofs of the
choir aisles were carried by semicircular arches.

In Ireland and in Scotland there are certain little primitive
stone buildings, of which the best known is the so-called oratory

Fig. 36.   Interior view of the farm-house near Strata Florida Abbey, of which the plan
is shown in Fig. 35.   The ridge tree is carried both by principal rafters and by an
upright post.   All the timber was unsquared.

of Gallerus. They are rectangular in plan and somewhat like an
inverted boat in appearance. In construction they are pointed
barrel vaults, resting on the ground, but with the stones laid in
horizontal courses, instead of in radiating courses as in a true arch.
They are apparently derived from the application to a rectangular
plan of the method of roofing the circular or 'beehive' huts, and
the upper part of the roof is formed by slabs of stone laid across,
copying the slab which closed the top of the beehive huts.   This

---

[1] 'Observations on the Origin of Gothic Architecture' in *Archæologia*, XVII (1814),
pp. 1 to 29.

method of building was adopted for the roofs of Irish Romanesque buildings with vertical walls, and also for the strange South Pembrokeshire churches of the twelfth and thirteenth centuries. From South Pembrokeshire the Anglo-Normans had set sail for the conquest of Ireland, after they had conquered South Wales, and so there is some possibility that the English became acquainted with the pointed barrel vault in that way. It must be said that one or two of the oratories are shaped internally like a reversed ogee and the roofs of some of the later buildings are somewhat like a Mansard roof, both of which recall to mind certain forms of crucks ; it is possible that the form of the oratories was derived from boat-shaped buildings which have perished, as there is no evidence for crucks in Ireland as far as the writer has been able to ascertain. The evidence in favour of the derivation of the pointed arch from other sources is strong, and must be preferred to theories of its wooden origin.

We have now to consider the more advanced methods of building by which the crucks were succeeded.

The old builders did not distinguish between the crucks and the principal rafters. It has been shown that the crucks were known as siles and levers and that the principal rafters are known by the same names, and the writer finds the same confusion in Wales where the old principal rafters are called 'cwpl' like the crucks.

CHAPTER VI

FULLY DEVELOPED TIMBER BUILDINGS

Anglo-Norman roofs and their descendants—Re-use of timber in building—
Post and truss buildings—Their use in the sixteenth century—Roof trusses
and their principal varieties.

Viollet-le-Duc expressed his surprise at the frequent use of
the tie-beam as the support for a short post, carrying the
ridge-beam, in the roofs of the Anglo-Norman churches[1]: this,
however, was only another device of the fertile and practical
English mind, to free the centre of an apartment from inconvenient supports. It was developed from the old king-post
which reached to the floor, by shortening the post and carrying
it, and its load, on to a cross beam and so on to the walls. The
development of such Anglo-Norman roofs is difficult to trace, as
so few remain from this early period, and in the thirteenth century
they were succeeded by the entirely different 'close couple' roofs.
If the shortened king-post were propped at the sides by raking
pieces, as was done in the Middle Ages, and these were made to
carry the purlins, then the type of roof truss was produced which
we still use. As already stated, it is unlikely that any form of
construction has only one line of descent and the ancestors of this
line were probably lost in the destruction of the smaller Anglo-
Norman halls and manor houses. In Fig. 37 a truss in a gable
end of a barn at Shire Green, Sheffield, is shown, in which the
king-post is propped in the above manner. The trusses inside
the building are similar, and the king-posts and their struts stand
free, but their position is not suitable for photography.

Strong stone walls, such as those of the Anglo-Norman
churches were rare and only used for the most important buildings.

[1] *Dictionnaire Raisonné, s.v.* Charpente, III, p. 25.

Sometimes the carpenters, true to their Northern ancestry, seem even to have mistrusted the strength of the stone walls and put in timber wall-posts strong enough to carry the roof, independently of the stone walls, as was done in a large barn at Peterborough, now destroyed, where the stone walls were a mere casing, so that when they were removed in the demolition, the whole of the timber supports, and the great roof which they carried still remained standing[1]. In the recurring collapses of stone build-

Fig. 37.   Barn at Shire Green, Sheffield, in which
the transition from struts supporting the shortened
king-post to principal rafters is shown.

ings in the Middle Ages, the carpenters had ample justification for their faith in their own material; moreover, walls during the period were usually only of boards, or wattle and daub, or lath and plaster, and were, therefore, not strong enough to carry the roof. Timber wall-posts were necessary in such cases, and buildings formed in this way are named 'post and truss' buildings in this book. They are the later of the two great types of English timber buildings, the other and earlier being those buildings on crucks which have already been described.

[1] *Report of the Society for the Preservation of Ancient Buildings*, June 1892.

Although it was usual to use the same oak timber in successive rebuildings, the crucks in the Sheffield district show no signs of previous use: they were evidently new timber when they were framed up as crucks. This is not the case with the post and truss buildings, in the framing of which crucks have been very evidently reused. At a Derbyshire house built in the year 1658—Offerton Hall, in the vale of Hope—the principal rafters of the roof, visible in the attics, seem to be the reused crucks of an earlier and one-storied house. In some seventeenth century buildings at Langsett and Penistone, South Yorkshire, crucks from older buildings were used for tie-beams in post and truss buildings.

The posts of the buildings in the Sheffield district, and also in Kent and Surrey, and possibly in other counties are placed upon their heads, with the butt, or root end, at the top: the crucks on the contrary are placed naturally, as the trees grew. Sixty years ago, in a wooden fence which had been fixed for fourteen years, it was found that all the posts which were fixed in the same direction as the growth of the timber had rotted, but that those which were inverted had remained sound[1].

Another difference between the buildings on crucks and those constructed with posts and trusses is more important than it seems: in the former the wall-plate is carried by the tie-beam, while in the latter the tie-beam is laid upon the wall-plate.

Post and truss buildings of the fifteenth and of the seventeenth centuries are still numerous, but this kind of construction reached its height in the sixteenth century. The buildings then erected are of this kind wherever the necessary timber was obtainable and the number of examples which still remain is very large. All the most celebrated timber and plaster buildings of England are of this type, and generally they are of the sixteenth or early seventeenth century. Not only the richly decorated old halls of Cheshire, but the humble cottages over wide districts of our country are framed on a skeleton of this fashion under their wrought pargetting or their rough daub. In the south-eastern counties of England the post and truss method of construction is supreme in timber built buildings[2], and in South Yorkshire

---

[1] *Builder*, XII (1854), p. 468.

[2] As far as the writer knows, no examples of any framed-timber construction older than the post and truss method are to be found in the counties shown by widely-hatched lines on the map (Fig. 12). Greensted church is hardly to be called an example of framed construction.

many examples remain, although they are often incomplete, having suffered from the inroads of new fashions.   The South Yorkshire examples belong to one type, which came into the district fully developed, and must therefore have been evolved elsewhere.   The widespread use of post and truss buildings coincides with the rise under the Tudors of the middle or bourgeois class whose ideals are peace and prosperity.   The low and one-storied buildings on crucks were abandoned for those on storey posts, of which the height was only limited by practical con-siderations.   At South Hinksey, near Oxford, cruck buildings and post and truss buildings stand together and may be compared, and the advantages of the latter must have been readily apparent to the practical English mind[1].

With the post and truss building came the end of the supremacy in buildings of the wright, the craftsman who had constructed the homes of Englishmen for long centuries, for in the seventeenth century the growth of the navy and the con-sumption of timber for fuel made wood more costly as a building material than some others which were as readily obtainable and which naturally took its place.

Moxon, the seventeenth century writer of the *Mechanick Exer-cises*, says nothing about crucks.   His timber buildings have posts, and only the principal ones, at the four corners, are to run up the whole height of the building in one piece.   They are to be eight inches broad and six inches thick, whereas in some old South Yorkshire barns they are fifteen inches by ten inches.   Moxon says the posts when set up are to be kept temporarily upright by tacking a deal board to them, and to the floor, diagonally[2].

The examples which remain show that the typical post and truss building in South Yorkshire was constructed in the following manner.   On stones at the ground level two upright posts were placed opposite to each other, and the width of the building apart.   They were set with their butts—that is, their root ends—upwards, and the posts became thicker towards the top at their inner side: on the top of the post two rebates formed a tenon

---

[1] Robert Reyce in his *Breviary of Suffolk*, written in 1618, says the wastage of timber has enforced 'a new kind of compacting, vniting, coupling, framing and building with almost half the timber which was wont to bee vsed, and far stronger as the workemen stick nott to affirme, butt the truth thereof is nott yett found outt soe,' p. 50.

[2] He describes the framing and rearing of a post and truss building (pp. 135 et seq. with an illustration, Plate 11, facing p. 147).

running in the direction of what was intended to be the outer wall. Another rebate was made at the top of the post on the outer side, and two smaller rebates formed another tenon running in the direction of the width of the building. The wall-plate was fixed in this lower rebate, and upon the tenon, which

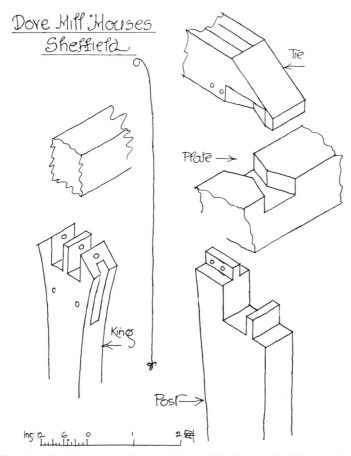

Fig. 38. The joints of the timbers of Dove Hill Houses, Sheffield, a typical post and truss building. The vertical positions are correct.

prevented longitudinal slipping. The cross tie-beam was laid on the top of the wall-plate, and the tie, in its turn, was held from longitudinal slipping by the upper tenon. These formed the main timbers and they were strengthened by means of struts, the principle being the same as in the wind braces of the cruck buildings, which are raking struts. If there was an upper floor to be carried, mortices were cut laterally on the inner sides of the posts, and beams were fitted into these at the level of the first floor, running in the same direction as the wall-plate : the beams carried the ends of the joists of the first floor. Another beam ran across the building at the same height, which served to hold together the posts at opposite sides of the building. Mortices were then cut under the wall-plate, under the tie-beam, and under

Fig. 39. Dove Hill Houses, Sheffield, in process of demolition. The post and truss construction had been concealed by later walls of stone, and was only revealed when the stone slates were stripped from the roofs and the outer walls pulled down.

the beams at the level of the first floor, and further mortices were also cut in the main upright posts—on three sides of the intermediate posts and on the two inner sides of the corner posts. The elaborate numbering of the mortice holes in many ordinary buildings is an indication that the mortice holes were cut and the timbers framed together provisionally on the ground. The mortice holes secured raking struts or braces, which helped to hold the timbers together and to decrease the unsupported bearing of the main horizontal timbers. The various joints of an ordinary post and truss building are shown separated, but in their correct vertical positions, in Fig. 38 from Dove Hill Houses, Endcliffe, Sheffield, which were destroyed a few years ago and measured at the time by the writer. The timberwork had been cased in walls for a hundred years or so, and so had been well preserved. It will be readily noticed that these joints are much more elaborate than the simple notchings and halvings of the cruck buildings. The Dove Hill houses in process of demolition are shown in Fig. 39 and their

Fig. 40. Ruined cottage near Firth Park, Sheffield, showing typical post and truss construction.

general appearance may be compared with the typical buildings on crucks shown in earlier illustrations. Another Sheffield cottage of the same type is shown in Fig. 40.

The roof trusses have tie-beams and principal rafters, but their other parts vary, and may be grouped into three principal types. In the first type, as at Dove Hill houses, there are the so-called king-posts of modern treatises on building construction. At Dove Hill the king-post had two clefts in the top which gave it the appearance of a gigantic three-pronged wooden fork. The ridge piece was placed upright in section and not diagonally, and a mortice on its under side was filled with the centre 'prong' of the 'fork': this prevented horizontal slipping and as the ridge was fitted between the two outer 'prongs,' side slipping was also prevented.   The ridge was also supported, in its long bearing, by struts from the king-post.   Moxon, in the year 1679, indicated the king piece or joggle piece (the modern king-post) as the truss and Sir Christopher Wren applied the term crown-post to both king-post and queen-post.   At Dove Hill the principal rafter butted against a 'shoulder' on the king-post, and from the underside of the principal rafters, at about their centre, struts ran to a 'knee' on the king-post.   The roof at Dove Hill, apparently of the Eliza-bethan period, was a fully developed piece of carpentry, for trusses with king-posts and struts are still the standard form, and have superseded the other types.   Thus it may be said, that the oldest method of carrying the ridge-piece is still in use, though the king-post no longer runs down to the floor but is carried by the tie-beam instead.

The second type of roof-truss had a horizontal straining beam, or collar, about halfway between the tie-beam and the ridge. This collar beam had various names, some of which survive in the dialects.   In Lincolnshire and 'other counties,' the collar is known as a span piece[1], and to Moxon, in the year 1679, it was known by the various names of top beam, collar beam, strut beam and wind beam[2].   A writer in *Notes and Queries*, 9th Series, VI, speaking of the contemporary roofs, says the 'whim beam (or... win-beam), is dovetailed into the main spars, about halfway between the ridge board and the baulk tie, which rests on the wall-plates, thus holding the spars firmly in position[3].'

The third type of roof truss was formed by propping the principal rafter by raking struts from the tie-beam, and at right angles to the rafter, generally at the point where it carried the purlin.   These struts were usually so placed that there was a space on the beam between their feet.   The builders did not always make their trusses uniform in design, and in a building at Falthwaite, South Yorkshire, one truss has two vertical posts, while another has a pair of raking struts, and two small vertical posts, in addition, between the struts and the wall.

[1] *Dictionary of the Architectural Publication Society, s.v.* Span Piece.
[2] *Mechanick Exercises*, Carpentry, Plate 11.
[3] p. 289.

The combination of the three principal types produced many variations, especially in the larger buildings of the seventeenth century. Thus, in the barn at Romely Hall, in North Derbyshire, there are raking struts from the tie-beam to the principal rafters, and the rafters are connected by a collar above the struts[1]. Another variety, somewhat resembling that at Falthwaite, had two small posts, which have been called 'queen-posts,' underneath the purlins, and in addition a king-post supporting the ridge.

[1] A somewhat similar truss, with the addition of a king-post (called 'king piece' or 'joggle piece') is shown in *Mechanick Exercises*, Carpentry, Plate 11.

CHAPTER VII

DETAILS OF TIMBER BUILDINGS

Post and truss buildings with aisles—Roofs without ridge-pieces—The old
German sparrendach—Rearing and the rearing supper—Movable buildings—The reason, pan or wall-plate—Purlins and rafters.

Unlike the buildings on crucks, the post and truss buildings
could be conveniently widened by the addition of side aisles. In
order to effect this, lower posts were placed at such a distance away
from the principal posts as to give the required width to the
building. The principal posts then corresponded to the columns of
a church arcade, and the smaller posts took the place of the aisle
walls. The roof was formed by continuing the principal rafter of the
central aisle down to the heads of the outer posts, and a wall-plate
and a tie-beam were framed together upon them, as in a single aisle
building, and the strutting and bracing also agreed. The tie-
beam of the side aisles was run against the principal post and was
morticed into it. In Fig. 41 the central aisle of the great barn at
Gunthwaite, near Penistone, South Yorkshire, is shown. In date
these timber barns with aisles extend to the fourteenth century, of
which period there are examples still existing in England. Whitaker[1],
the historian of much of the North of England, thought that they
succeeded the barns built on crucks, and that is the case in South
Yorkshire, where they are contemporary with the post and truss
buildings without aisles. Where a larger building was needed the
aisles of the post and truss building provided a method of lateral
extension, in place of the unavoidable lengthening of the building
on crucks.

In the time of King Henry VIII there was a kind of post and
truss building constructed in South Yorkshire which differed from
the normal type, already described. The principal difference is in

[1] *History of Whalley*, 1818, p. 500.

the absence of the ridge-tree, the common rafters being arranged in pairs and halved and pegged together at their upper ends.   The tops of the principal rafters were framed together in the same manner as the common rafters, and the purlins were framed into their sides, which threw the feet of the ordinary rafters on to the wall, and in order to carry the roofing beyond the walls the pitch of the roof was broken by triangular pieces, pegged on to the feet

Fig. 41.   Interior of Barn, Gunthwaite, near Penistone, showing the usual method of framing the timbers of a post and truss building with aisles.

of the rafters.   The roof of the Long Gallery at Sheffield Manor House shown in Fig. 42, constructed in this manner, also has wall-plates of the full width of the post underneath them.

The pieces at the feet of the rafters had various names, such as firring or shreading or sprocketing : the two former were used by Moxon[1] but the latter has now become the standard term.

According to K. Rhamm[2], the old roofs of the Low German (Platt Deutsch) speech lands were constructed with 'sparren,' that is rafters, without a ridge-piece, as in the South Yorkshire roofs of the early sixteenth century, and it is therefore possible that these

[1] *Mechanick Exercises*, Carpentry, Plate 11.
[2] *Urzeitliche Bauernhofe*, Chap. x.

were due to the example of the Low Countries or the Hansa
Towns. The South Yorkshire carpenters soon went back to the
use of the ridge-tree, ancient and time-honoured in English roof
construction.

In South Pembrokeshire, where Flemings settled, the writer has
seen a humble variety of 'sparrendach.' In it the rafters were
placed about four or five feet apart, with a collar to each pair of
rafters but no ridge-piece: the thatching laths ran from one pair of
rafters to another. The building was a plain old cottage of one
storey and it was not easy to assign a correct date.

Fig. 42. The Long Gallery at Sheffield Manor House, showing a Tudor roof without
a ridge-piece. The rafters were halved and pegged together at their upper ends.

In a roof with a ridge-piece, it is only necessary for the rafters
to be straight in one plane, that of the roof: but in close couple
roofs they must be straight in two planes. From this it may be
argued again that the roof with a ridge-piece is the more primitive.

The so-called close couple roof was formed of common rafters
braced together as the builder thought suitable, and fixed apart at
a distance usual with common rafters. Each couple was independent, and there was neither purlin nor ridge-tree. The close
couple roof was also called a cradle roof, and the term 'couple
close' is used in the North of England 'for a pair of rafters
framed together with a tie fixed at their feet, or with a collar

beam[1].' Norman roofs had tie-beams, and in the close couple roofs the carpenters at first adhered to the use of a tie-beam, and then logically abandoned it[2]. The thirteenth century close couple roof was an innovation, probably founded on French precedent, like the plan of Westminster Abbey, and in due course the English carpenter returned to the construction of roofs with ridge-pieces.

Some English domestic buildings of the thirteenth century have close couple roofs, without a ridge-tree as is usual, and with a tie-beam. This supports a short king-post which does not reach to the ridge, but carries a longitudinal beam which supports the short binding collars of the close couples. A reference to such a book as Brandon's *Parish Churches* will show that such a beam was not used in churches, and was therefore not structurally necessary for the support of the short collars. In Switzerland there are houses with ridge beams and with another horizontal beam underneath the ridge beam, which does not serve any structural purpose, and is known as the 'katzebaum,' that is, cat beam[3]: its original use was, perhaps, as a perch for the house owl, which in early times took the place of the cat in Central Europe. It may have had the same origin in England as in Central Europe.

The timbers of the crucks and also of the more elaborate posts and trusses were framed together on the ground and then hauled up into position. This was known as the 'rearing.' The timbers were so heavy that it was necessary to call in the neighbours to help and it was usual to feast them afterwards for their pains, thus, in the year 1420 the churchwardens of St Michael's, Bath, paid twopence for drink at the 'rerying' of a house at Alford[4]. In the year 1530 Palsgrave wrote, 'My house is framed all redye (charpente): it wanteth but setting up.'

It was necessary to call in the public to rear even in such a town as sixteenth century Sheffield, and a householder there in the year 1575 'paid £2. 6s. 8d. "for meat and drink that day the house was reared," a large sum considering the rate of wages at that period[5].'

[1] *Dictionary of the Architectural Publication Society, s.v.* Cradle Roof and Couple.
[2] R. and J. A. Brandon, *An Analysis of Gothick Architecture*, 1858, II, pp. 91, 92.
[3] E. Gladbach, *Schweizer Holzstyl*, Figs. 28 and 29.
[4] *Proceedings of the Somersetshire Archæological and Natural History Society*, 1877–80.
[5] S. O. Addy, *Evolution of the English House*, p. 107.

In the year 1545 a law was made against certain novel outrages. 'The secret burnynge of frames of tymber prepared and made by the owners thereof, redy to be sett up, and edified for houses[1].' Such outrages evidently imply some change in method or in the conditions of labour, and they coincide in date with the wider use of post and truss construction, and its more elaborate framing.

When the English crossed the Atlantic to found a New England beyond the ocean, they took with them their custom of framing the timber together before fixing. At Plymouth, Massachusetts, in the year 1639, Governor Bradford recorded that 'Thomas Starr...hath sould unto Andrew Hellot one frame of a house, with a chimney, to be set up and thacked in Yarmouth in the place appointed[2].'

The little flag which decorates the roof of a modern house when it is ready for the slater is a survival from the days when the setting-up of the house was the 'rearing' of crucks or other heavy timbers by the neighbours, and a reminder of a still more remote time when the house was a movable structure.

On some South Yorkshire farms there are fields called Husteds, which indicates that they are the former site of the farm house, for building was then a cheaper and more simple affair than it is to-day. Not only were the sites of buildings changed more readily in early times, but the buildings themselves were moved. The Ancient Laws of Wales say 'Let the spars and posts be cut even with the ground, and let him depart with his house...for the land is no worse for transporting the house across it, so that corn, or hay, or dike be not damaged.' Bracton, in his *De Legibus Angliæ*, distinguishes between a wooden house 'whether it is attached to the ground or not.' A clause of the Assize of Clarendon, in the year 1166, says that any one who shall receive certain heretics shall have his house carried outside the town (*extra villam*) and burnt[3].

The timber house could be moved bodily. Wm Horman in his *Vulgaria*, published at London in the year 1519, spoke of a house being removed with 'trocles' and 'slyddis,' and says that "it shal be done with myght of men and oxen[4]."

Mr Bernhard Olsen has kindly supplied the writer with some

---

[1] S. O. Addy, *Evolution of the English House*, p. 108.

[2] *Records of the Colony of New Plymouth in New England*. In the same year, in an agreement with a carpenter, John Mynnard, for building a prison, it was agreed that help should be allowed him to rear the framing, at the colony's charge (1, p. 115).

[3] Stubbs, *Select Charters*, p. 173.

[4] p. 244.

Continental examples of the moving of buildings, covering a wide range in time from the days of the Norsemen to the seventeenth century. He writes that in the Middle Ages the landlords let the ground on one year's lease to the 'colonus' (corresponding to the English 'villanus' or 'cottar'). If he wished to leave, and was unable to sell his house to his successor, he took it away with him. Many examples of the moving of buildings in Germany are known[1].

Useful information may be obtained from the old glossaries and dictionaries as to the parts of the roof, as it may for the other parts of the building.

We have seen that the Mediæval Latin word 'laquear,' in a variety of spellings, was used for the ridge beam, and the glossaries show that it was also applied to the other timbers which ran longitudinally in the direction of the length of the building. The word is translated 'raesn' in the eleventh century; as both 'fierst' and 'raesn' in the twelfth; and as 'lace' and 'post bande' in the fifteenth. In 1338 an item occurs 'for six pieces of timber bought for the rasen of that house[2],' and the *English Dialect Dictionary* shows that 'reason' still means wall-plate in the dialect of the southern and eastern counties of England. In the seventeenth century Ray wrote that the pan was variously called 'rasen,' or 'resen,' or 'resening' in timber buildings in the South. Gerbier, an architect of the same century, defined the reason piece as the wall-plate. The word 'pan' does not occur in the Wright-Wülcker *Vocabularies*, but in the *Catholicon Anglicum* the 'panne of a howse' is translated by 'panna.' The earliest quotation for the word in the *Oxford Dictionary* is from the Close Rolls of the year 1225: 'Palnas, postes, trabes, and cheuerones,' that is plates, posts, beams, and rafters. Cotgrave defined 'panne de bois' as the piece of timber that sustains a gutter between the fronts of two houses, which is the wall-plate, as such houses generally had their gable ends to the street. The word is still in use in the dialects of Scotland and the North of England. In Lancashire and Scotland it is used for the purlins, but in South Yorkshire it is restricted to the wall-plate. The word 'pan' is of Norman-French origin, and it is strange that a Norman-French word should be used in the North of England and a good

---

[1] R. Henning, *Deutsche Haus*, p. 164.
[2] *Oxford English Dictionary*, *s.v.* Reason.

English word be retained in the South, as the contrary is usual. Possibly the ruder North still retained roofs running down to a sill piece on the ground after the buildings in the South had perpendicular walls.

In the fifteenth century *Catholicon Anglicum* the word 'laquearium' is translated 'a bande of a howse': this word band was also applied to the purlins, and in the *Promptorium Parvulorum* of the same century 'laquearium' is translated 'Lace of a Howserofe.' The purlins have a variety of names in the dialects, which implies that they are not as ancient as the ridge, and probably that they are of varied ancestry. In Cheshire and in Shropshire they are 'side rezors,' or 'rizzars,' probably a form of the old English 'raesn.'

In Middle English the purlins were also called ribs, a name which is found in building accounts, and they are called 'rib-resenes' in a nominale, of about 1340. In the dialect of Ayrshire the purlins are called either pans or ribs. Another old name for the purlin, 'waver' or 'weaver,' is still in use in the dialects of Yorkshire and Lancashire in the form of 'side weaver.'

'Laces, or binding beams' and 'purlaces' occur in Holme's *Academy of Armory*, of the year 1688, in the enumeration of the 'several pieces of timber belonging to a wood house[1].' The purlins are called 'sleepers' by Moxon, who also defines a 'plate' as a piece of timber on which considerable weight is placed, as 'ground plate,' 'window plate[2].' The earliest use of the word in this sense given in the *Oxford Dictionary* is in a quotation dating from the year 1449, which says 'The plates of the same hous shullen be in brede x inchis and in thiknes viii inches.' Purlins are 'side pieces' in the dialect of Northamptonshire, and in that of South Yorkshire they are 'side trees.'

The timbers which run from the ridge to the plate, or, in primitive buildings to the ground, and of which the purpose is to carry the roofing materials, are called 'spars' in the North of England, and 'rafters' in the South. In Scotland they are 'cabers' or 'kaibers.' Spars occur with ribs, and with firsts (ridge beams), in *Durham Halmote Rolls[3]* of the year 1378, and in the *Catholicon Anglicum* the Latin words 'tignus' and 'tigillum' are translated 'sparre.' In Anglo-Saxon times 'tignus' is translated 'rafter'

---

[1] p. 450. Many of the building terms given by Randle Holme, seem to have been taken from Moxon's *Mechanick Exercises*.

[2] *Mechanick Exercises*, pp. 170 and 172.   [3] I, p. 149.

(Wright-Wülcker *Vocabularies*). The Norman-French word for rafter was 'chevron' and the records show that Henry III was constantly making grants of them to his subjects, and it has been pointed out that when the king made presents of chevrons, they were always in multiples of five[1]. Originally the rafters were natural poles, then split poles were used, probably to give a better hold for the laths, then these were roughly squared, and so old rafters came to be laid on their flat sides, as the builders knew nothing of mathematical formulae of the strengths of materials. In South Yorkshire old roofs with rafters of split poles still remain and old roofs constructed with natural round poles also; while in Surrey the rafters of old roofs were either pit sawn or squared with the adze[2].

The operation of joining the beams together is called 'knitting the siles,' and the fifteenth century *Promptorium Parvulorum* gives 'knyttynge or ioynynge or rabatynge to-gedr of ii bordys or otherlike.' The principal rafters were known as 'sile blades' in the West of Scotland and as 'sile trees' in Cumberland, and the social gathering at which the principal timbers of a roof were reared was called the 'sile raising' in Westmorland. The principal rafter is called the 'back' in Lancashire[3].

The beam which runs across the building was generally called 'trabs' in Mediæval Latin, although both 'trabs' and beam have a somewhat wide meaning, and the word is also translated 'balk' in the fifteenth century glossaries. 'Trabecula' (literally a little beam), with the usual varieties of spelling, is translated 'bind balk' and 'wind beam' in the same glossaries so that the builders of the period believed that the straining piece or collar, midway between ridge and tie, served both as a tie and as a strut.

In Germany the cross beams are known as 'hahnebalken,' that is, fowl beams[4], and in England a similar beam was known as a 'perch' ('pertica')[5]. The Elizabethan bishop Hall, in his *Satires*, described a poor Englishman's cottage which had 'his swine beneath, his pullen o'er the beam[6].'

---

[1] G. J. Turner, *Select Pleas of the Forest* (Selden Soc.), p. 140.
[2] W. Curtis Green, *Old Cottages and Farm Houses in Surrey*, p. 27.
[3] *Dictionary of the Architectural Publication Society*, *s.v.* Principal Rafter.
[4] J. and W. Grimm, *Deutsches Wörterbuch* ('der oberste Querbalken unter der Dachfirst, wo der Haushahn seinen nächtlichen Sitz zu nehmen pflegte').
[5] S. O. Addy, *Evolution of the English House*, p. 50.
[6] Bk v, S. 1.

There is a common form of roof in which the ends of the building, instead of being upright gables, are sloped back from the wall-plate to form a roof which is triangular in shape, when the side roofs meet at a ridge as is usual. It is known as a hipped roof. In the Teutonic and Slav lands of the Continent such roofs are usual and evidently of early origin, but there appears to be no evidence for their age in this country. In the North they seem to be associated with the Renaissance, for in Scotland a hipped roof used to be known as an Italian roof. The rafters which run up the intersection of the planes of a hipped roof are called hip rafters and in the older examples in the South of England they do not meet at the ridge, but finish below it, leaving a small gablet. The same construction occurs in Germany, and in old houses the gablet was not filled in, but left as a hole for the house owl to go in and out. Its dialect name was 'ulenlok,' that is, 'owl hole[1].'

There is an intermediate form of hipped roof in which the gable wall is carried up for a part of the height, and the roof is then 'hipped back.' The original roof in the old barn at Falthwaite, South Yorkshire (Fig. 27), was probably constructed in this manner as is the cottage on crucks in Fig. 55. The triangular roof so formed is called 'pirk,' 'jirkin head,' 'kirkin head' and 'shread head' in the dialects. The derivations of these terms are unknown, but pirk is probably connected with the use of the word perch for the collar beam.

The earliest literary use of the word 'hip' and 'valley' appears to be by Moxon in 1677. He gave to the hip the names 'principal rafter' and 'sleeper,' also.

---

[1] K. Rhamm, *Urzeitliche Bauernhofe*, p. 219.

CHAPTER VIII

THE CARPENTER

The wright, or carpenter and his work—His history—The descent of the joiner
and the cabinet-maker—The carpenter's tools in the documents—Oak,
Baltic timber, mahogany.

The timber framework, which has been described in the
preceding chapters, was the work of the wright, a craftsman who
also attended to other woodwork such as the household and farm
utensils and furniture. In Aelfric's time, according to his *Colloquies*,
the wright did all work in wood that was required and occasionally
fashioned utensils from other materials: besides the heavier work
of house building, the making of tubs, buckets and vats came
within his province. Among the labours of the 'gerefa' or reeve,
before the Norman Conquest, was the supervision of the splitting
of timber with wedge and beetle during the time of great frosts,
together with the chopping of wood by the axe, and much building
was done in June and July. It was the duty of the 'gerefa' also
to see that the wright constructed tables and benches for use in
the house[1]. In the fifteenth century tale of *The Wright's Chaste
Wife*, the wright is said to work "hows, harowe, plowgh and other
werkes."

In the middle of the last century, wright was 'the general
designation in Scotland for all who work in wood, a common
carpenter[2].' Robert Manning of Bourne, glossing native English
words with Old French, early in the thirteenth century, equated
wright with carpenter. The Mediæval Latin word for wright was
'carpentarius,' the craftsman who made the 'carpentum' or wain
and in Anglo-Saxon glossaries 'carpentarius' was translated by
'waenwyrhta,' that is wain or cartwright, or by 'waengerefa,' that

---

[1] C. M. Andrews, *Old English Manor*, pp. 249 and 257.
[2] *Dictionary of the Architectural Publication Society*, *s.v.* Wright.

is wain reeve. In the fifteenth century the word was variously translated carpenter, cartwright, and wright. Probably a large number of carts were 'sleds' without wheels, such as were used in the streets of Bath, Bristol and Derby in the time of William and Mary according to the diary of Celia Fiennes, and are still used on farms in North Derbyshire. The wright who specialised in wheels was called a wheelwright and it is likely that the making of carts became a specialised industry before the construction of buildings and that the wright, as the professional woodworker, took to the construction of the buildings as they were developed beyond the constructive capacity of their owners.

The joiner was another craftsman in wood who occupied himself with building. At first he seems to have been what we should now call an upholsterer, who probably worked also at the frames of the furniture, for the numbers of the joiners, as makers of furniture, increased with that rise in the standard of culture in the sixteenth and seventeenth centuries which brought furniture into general use for the first time among the middle class. Such furniture is the 'joined' furniture of the wills and inventories of the time, and many of the joiners were foreigners. The joiners then began to take part in the fitting up of buildings and in the seventeenth century there were contentions between them and the carpenters as to their proper work, analogous to recent misunderstandings between the plumbers and the heating engineers. In each case, the cause was the irritation of the members of an ancient craft at what they considered to be an intrusion on their work by one that was comparatively new. In the year 1632 the differences between the work of the carpenter and that of the joiner were similar to those of the present day, and Moxon wrote 'Joynery is an Art Manual, whereby several Pieces of Wood are so fitted and joined together by straight Lines, Squares, Miters, or any Bevel, that they shall seem one intire Piece[1].' In the year 1672 the carpenters complained that 'without question the Joyners trade before their incorporacon was chiefly to make and sell joyned ware, as bedsteds, tables, chaires, stooles, etc., and to joyne and ceele only, and not to doe any other worke about building of houses but what the Carpenters imployed them in (wich they did for expedicon only), themselves generally doeing those workes they imployed them in.' Since then they have composed their quarrels and the two trades

---

[1] *Mechanick Exercises*, p. 59.

are now more or less united.   Through the centuries, the carpenter
has kept the trade of a general worker in wood : the village 'joiner
and wheelwright' of the present day does the same kind of work
as did his predecessor of a thousand years ago, and he has kept
abreast of the times to some extent by adding joinery and simple
cabinet making.   There is an old Welsh proverb that the mason
gets worse and the carpenter better.   Moxon wisely says ' He that
knows how to work curiously, may, when he lists, work slightly :
when as they that are taught to work more roughly, do with greater
difficulty perform more curious and nice work.'

The late Mr Frederic Seebohm considered that the carpenter
was an official of the free village community, and held his
holding free in return for his obligation to repair the woodwork
of the ploughs and harrows[1].   If this ever were so, his status
had become lower at the time of our earliest knowledge of the
manor.   Probably before the Conquest the millard, shoemaker,
smith, and wright were already recognised as distinct craftsmen :
but all others, such as brewers, weavers, etc., were, and continued
to be for a considerable time, merely household servants[2].   As
such they were slaves, and in a grant by William the Conqueror
to Crowland Abbey there is an oft-quoted passage which relates
to the grant of slaves with the land, and these include the smith
and the wright.   Each slave was transferred with his offspring,
his goods and cattle, which he had in the said vill and in its fields
and marsh.

In a survey of the possessions of the Abbey of Glastonbury
in the year 1189, an improvement is shown, for the carpenters
paid rent for their holdings : thus, at Brent Marsh in Somerset, one
Robert held one 'ferdel' for three pence, and made and repaired
carts and wooden utensils and harrows and houses according to the
custom of the lord of the manor[3].   In another village, Shapwick,
one Henry held five acres for fifteen pence without service, and
this was a further improvement on the conditions at Brent Marsh[4].
At the date of the compilation of this Glastonbury 'Book' the
carpenter seems to have been ceasing to be the servant of the
community and its lord.   Then as time went on, the carpenter was

---

[1] *English Village Community*, 4th edition, p. 70.
[2] C. M. Andrews, *Old English Manor*, p. 237.
[3] *Liber Henrici de Soliaco, Abbatis Glaston*, p. 66.
[4] Ibid. p. 54.

paid for his labour on the materials which his employer provided. Professor Thorold Rogers[1] found smiths and carpenters very rarely mentioned in the rent rolls of the manors. Later, the carpenter provided and was paid for both the materials and his labour. Rogers found that the earliest capitalist artisans who dealt in finished goods lived in the eastern counties and London, and that their transactions were with opulent individuals or with wealthy corporations. This is another instance of the influence of the Continent upon English civilisation. In the fifteenth century the carpenter began to employ other carpenters to work for him, either on his own materials or on those provided by the employer, and by the seventeenth century there was a possibility that the successor of the former servant of the community or its lord might become a capitalist employer. These three, the carpenter providing his own labour only, the carpenter providing his own labour and the materials in addition, and the capitalist employer with other carpenters working for him, have all continued to the present day, as each is of use according to the kind of work to be done.

As recently as the year 1792, in the parish of Naseby, in Northamptonshire, which at the time was still cultivated under the old common-field system, 'The farmers, twenty-one in number, who keep teams, support constantly four working blacksmiths, two working wheelwrights, besides carpenters,' and in addition workmen of other trades unconnected with building[2].

As the carpenter held land, he must have done a little farming, and this seems to have been usual with all the village craftsmen. Rogers found that the employers of such a craftsman as the carpenter occasionally purchased agricultural produce from him or from his wife. One of the carpenters in the Glastonbury 'Book,' a certain William, held five acres and a mill at Shapwick for three shillings ('soldi'), but ought, the surveyor noted, to have paid five shillings[3]. In the Burton Chartulary, one Thorold, another carpenter who held a mill, had to do all the works of the church which pertained to his office, both of wood and lead. The latter was too expensive a metal to be used in the ordinary buildings described in this history, except in the neighbourhood of lead

---

[1] *Six Centuries of Work and Wages*, pp. 46, 179 and 338.
[2] Rev. John Mastin, *History and Antiquities of Naseby*, p. 15.
[3] *Liber Henrici de Soliaco, Abbatis Glaston*, p. 54.

mines, and in this instance the carpenter Thorold, in the absence of a plumber, did the necessary leadwork at the church. In the year 1805 it was observed that there were very few cottages in the county of Westmorland, as the labourers and mechanics generally resided in a small farmhouse and occupied more or less land[1]. In out-of-the-way country places it is still possible to find men who turn their hands to several occupations: they are relics, flotsam and jetsam, from an earlier stage of culture when industry was not specialised as it is to-day.

A specialised industry, to be satisfactory to the persons engaged in it, must provide constant employment: but at the present time industries are highly specialised without providing continuous employment for those engaged in them. In the Middle Ages, the holding of land by the craftsman provided him with a useful by-industry to add to his regular means of sustenance.

Delightful pictures have been drawn of the mediæval craftsman, of his love for art and of the happy and stimulating conditions under which he worked; but these do not receive support from the somewhat scanty evidence provided in the literature of the period. In the fifteenth century story of *The Wright's Chaste Wife*, the wright worked with a beautiful garland of white roses hung about his neck, but the wright's decoration was connected with the behaviour of his wife, and not with design or craftsmanship; its strangeness is shown by the openly expressed surprise of those who saw him.

There is a curious fifteenth century poem, *The Debate of the Carpenter's Tools*, in which the tools discuss their work and their employer after the manner of men, but his drinking powers and the amount which he spends upon liquor seem to be the characteristics of the carpenter which most interest his tools. His wife is made to complain that

> He wylle spend more in an owre
> Than thou and I canne gete in fowre.

She regrets that she ever married him and lays the blame on the parson for making them man and wife.

The attitude of the tools with respect to their work is shown in the following appeal of the belt:

---

[1] J. Bailey and G. Culley, *General View of the State of Agriculture in Northumberland*, p. 301.

Mayster, wyrke no oute off resone,
The dey is vary long of seson,
Smale strokes late us hake,
And soun tyme late us ees our bake.

Then the groping iron says :

Master, wylle ye well done?
Late us not wyrke, to we swet,
Ffore cachyng of over gret hete,
Ffore we may happe after cold to take,
Than on stroke may we no hake.
Than bespake the whetstone,
And seyd, Mayster, we wylle go home,
Ffore fast it drawe unto the nyght,
Our soper by this I wote is dyght.

The other tools agree, saying :

That is gode counesylle :
The crow, the pleyn, and the squyre,
Says we have arnyd wele our hyre :
And thus with fraudes and falsyd
Is many trew man deseyvid.

As far as this contemporary literature reveals the mind of the craftsman, it shows that in the fifteenth century, the period of splendid roofs and glorious screenwork, the carpenter was thinking as he worked, not of Art and Beauty, but of his wife, of his wages, of his beer and of the time when, his work done, he could lay down his tools and be off home to supper.

In this *Debate of the Carpenter's Tools* twenty-seven different tools took part. The full list is interesting as it probably includes all the tools then used ; they were : 'twybylle,' 'shype axe,' wimble, belt, groping iron, compass, saw, whetstone, file, adze, chisel, line and chalk, pricking knife, 'persore,' 'skantyllon,' crow, rule, plane, broad axe, 'twyvete,' 'polyff,' 'wyndas,' rule stone, gouge, 'gabulle rope,' 'draught nayle,' and square.

On the 19th September, 1597, about a century after the date of the *Debate*, the Corporation of Leicester caused an inventory to be made of the goods of one Will. Hobby, a wright, which was printed by Miss Bateson in the third volume of her *Records of the Borough of Leicester*. The yard attached to his shop was occupied by wood for carts, etc., and in the shop were two 'blocke sawes,' a whipsaw, seven axes, three hatchets, three adzes, six 'beerezees,' one great

hammer and two little ones, two pairs of pincers, an 'yron dogge,'
four spoke shaves, twenty-one augers, and other 'trincketes,' worth
altogether thirty-three shillings and four pence. This gives us an
indication of the stock of tools kept by a provincial carpenter in
the days of Queen Elizabeth.

Earlier than this, in the first half of the thirteenth century,
John de Garlande had written that timbers were wrought with
'hachet,' 'brode axe,' 'twybyl,' axe, wimble, wedges and pins, celt,
plane, mason's line, with 'reule' and 'squyre' and with 'hevy
plomet.'

Throughout the Middle Ages, the axe was the carpenter's tool
par excellence. There is an illustration[1] in a thirteenth century life
of Edward the Confessor, which shows three craftsmen taking
their instructions from the king: first a master mason in official
master's cap, wearing gloves, and carrying a long, levelling straight-
edge: kneeling before him is the carpenter with a coif on his head
and carrying an axe, and behind the master is another mason with
a stone axe.

The old glossaries give some indication of the former importance
of the axe: thus 'lignum,' that is building timber, is translated as
hewn wood in vocabularies from the time of Aelfric to the twelfth
century. 'Dolatum,' 'incisum,' and 'planum' are all translated by
hewed in the tenth century, and 'dolatum,' alone, is translated by
'sniden' in vocabularies of the eighth and eleventh centuries, so
that to the compilers of the pre-Conquest glossaries, timber with a
finished surface was that which had been dressed by the axe, or
which had been through the saw pit.

To-day, in England, we hardly realise the skill and speed
with which an axe may be wielded. It is said that the carpenters of Dalcarlia and Norrland, in Sweden, 'require no other
tools than the axe and the auger, and despise the saw and plane
as contemptible innovations, fit only for those unskilful in the
handling of the nobler instruments: they will trim and square a
log forty feet long as true as if it had been cut in the saw mill, and
will dress it to a face that cannot be distinguished from planed
work[2].'

The old German instrument for smoothing timber, the broad

---

[1] Reproduced by Professor W. R. Lethaby in his *Westminster Abbey and the King's
Craftsmen*.
[2] Alex. Beazeley, *Transactions of the Royal Institute of British Architects*, 1882–3.

bladed axe, was still in use round Ratisbon in the middle of
the last century, and the plane is stated to have been 'but little
known' on the continent of Europe up to the early part of the
century[1]. The broad axe ('dolabrum'), the ordinary axe ('securis'),
and the adze ('ascia') are all named in glossaries of the time of
Aelfric. According to Moxon (A.D. 1677) the adze was 'most used
for taking off the irregularities on a framed work of a Floor, when
it is framed and pin'd together, and laid on its place[2].' It was also
used for dressing posts and the irregularities of framed work on a
ceiling, etc. Early axes are said to have had their heads welded.

The saw is another tool mentioned in Aelfric's glossary, and
was probably the two-handled form, which was worked by two
men, one above and one below. 'Serrula,' literally a little saw,
rendered by Aelfric as 'saga' (saw) and 'snide' (slicer), was
probably a single-handed saw, and 'serra' is translated also by
'snide' in a vocabulary of the eleventh century. There were four
kinds of saw in use in the Middle Ages[3], (1) the great pit saw,
(2) the frame saw or whippet saw, (3) the hand saw, and (4) the
twart saw.

In Moxon's time, besides the two-handed saw, there were the
'Whip-Saw, Hand-Saw, Frame or Bow Saw, Tennant-Saw, and
Compass-Saw.' He says 'Chuse those that are made of Steel, (for
some are made of Iron)[4].' The steel saws were ground but the iron
saws were only hammer hardened.

When stuff was to be sawn in the pit, it was first marked with
the chalk and line, and then fastened with wedges over the pit. If
the joiner had no pit he made 'shift with two high frames a little
more than Man high in its stead, (called great Trussels) with four
Legs, these Legs stand spreading outwards, that they may stand
the firmer: Over these two Trussels the Stuff is laid, and firmly
fastned that it shake not[5].'

The modern dictionaries of Anglo-Saxon give plane for the A.S.
words 'sceaba' and 'locer,' which are translations of 'runcina' in the
so-called *Corpus Glossary* of the eighth century. If they were planes

---

[1] *Dictionary of the Architectural Publication Society*, *s.vv.* Chip Axe and Plane.

[2] *Mechanick Exercises*, House Carpentry, p. 122.

[3] According to an anonymous writer in the *Builder*, 7th July, 1911, who based his
statement on illustrations in manuscripts, which generally show the construction of large
buildings where the tools and appliances would be of the best.

[4] *Mechanick Exercises*, p. 94.   [5] Ibid. p. 97.

at all, it is unlikely that they were so elaborate and finished as the planes of to-day : the first name seems to imply a shave, but planes were in general use by the end of the Middle Ages. In the article in the *Builder*, it is said, from the evidence of the illustrations, that the mediæval plane was like our trying plane, and that it had a big bead in place of a handle. In the fifteenth century *Catholicon Anglicum*, a plane is translated by 'instrumentum,' 'dolabrum,' 'leviga,' and 'planatorium,' and in the contemporary *Debate of the Carpenter's Tools*, the plane says that it ' schalle clens on every syde,' and the 'brode axe' says 'the pleyn my brother is : we two schall clence and make full pleyne.' By the time of Sir Christopher Wren the plane was fully developed, and its varieties, fore plane, jointer, smoothing plane, 'rabbet' plane, 'plow,' and moulding planes of various sorts were described by the contemporary Moxon in his *Mechanick Exercises*[1]. In the *Catholicon Anglicum* a 'mortas' is rendered 'castratura,' and in the *Promptorium Parvulorum* 'incastracio' is translated by 'growpynge,' while a 'growpynge' or 'graovynge yryn,' is 'runcina,' that is literally, a groper. The *Ortus Vocabulorum* defines 'runcina' as that tool of the wood-worker, graceful and recurved, by which boards are hollowed so that one may be connected with another—Anglice, a 'gryppynge yron' ('est quoddam artificium fabri lignarii gracile et recurvum, quo cavantur tabuli, et una alteri, connectatur'). Palsgrave calls a gouge ' formour or grubbyng yron,' and the Anglo-Saxon word was 'graep.' The *Promptorium* translates 'formowre' by 'scrofina,' and the *Catholicon* explains that a ' scrobe' is called 'scrofina' because it is the instrument with which the carpenter makes a groove (' quoddam instrumentum carpentarii, quia herendo scrobem faciat'). Moxon says that ' Formers are of several sizes' and are so-called ' because they are used before the Paring Chissel, even as the Fore Plane is used before the Smoothing Plane[2].'

Mortice holes were bored at each end with an auger or a brace, so that the mortices have circular ends in the timber of the old buildings in the Sheffield district : the remaining wood is said to have been chopped away with a twivil[3]. The word ' auger' occurs in English as early as the eighth century, that is before the coming of the Danes. In the Ancient Laws of Wales three boring instruments are mentioned, viz. the larger auger, the middle-sized

---

[1] pp. 61 et seq. [2] *Mechanick Exercises*, pp. 72 et seq.
[3] *Dictionary of the Architectural Publication Society*, s.v. Twivil.

auger, and the wimble. The modern Welsh word 'wymbyll' is
now translated by gimlet. Augers, therefore, were of various sizes
at an early date, and Mr G. J. Turner, in his *Select Pleas of the
Forest*, printed a Buckinghamshire agreement of 1286, in which
a certain John had a right to such thorns as could not be perforated
by an auger ('tarrera') that is called 'rest navegar.' 'Restwymbyll'
occurs in the year 1446 and 'restwomyll' in 1464: the word
means a twisted bill used with a rest, which was a trough of
wood or metal placed horizontally to support the tool when in use,
and ensure that the hole was accurately bored[1], as through the
trunk of a tree to form a pump or water conduit. When a house
at Sharrow, Sheffield, which had been built in the latter half of the
seventeenth century, was pulled down, it was found that the pipes
to the well were of trees bored through, from end to end. The
boring tools were known in Mediæval Latin as 'rotrum,' 'terebrum,'
and 'terebellum.' 'Foratorium' is given as an alternative of
'terebellum' in some Anglo-Saxon vocabularies, and is translated
as 'nafogar,' that is auger, meaning literally 'nave dart,' and
'heard hewe,' that is chisel. The word 'furfuraculum,' as an
alternative of 'terebrum,' or its diminutive 'terebellum,' occurs in
Aelfric's Glossary and is translated by 'thystru,' and centuries
later, in a vocabulary of the fifteenth century, by 'persour.' In the
Promptorium both 'persowre' and 'wymbyl' are translated
'terebellum,' and in the fifteenth century 'wimble' is given as an
equivalent of 'penetrale.' In the *Debate of the Carpenter's Tools*,
the wimble describes itself and its duties:

> Zys, zys, seyd the wymbylle,
> I ame as round as a thimble.
> My mayster's werke I wylle remembyre,
> I schall crepe fast into the tymbyre.

And the 'persore' says:

> Fast to runne into the wode
> And byte I schall with moth (mouth) full gode.

According to Moxon, the piercer in his time was a drill with a
bit, the auger was used for great holes and the gimlet for small:
'the pricker was vulgarly called the awl.' The term 'wimble' is
said to be still used by boat builders and wood shipwrights in the

---

[1] *Dictionary of the Architectural Publication Society*, *s.vv.* Restwomyll and Womble.

West and the Midlands of England and is applied to a brace with a fixed bit made entirely of wood[1].

The word 'chisel' does not occur in the Wright-Wülcker *Vocabularies* before the fifteenth century, when, in the form of 'schesel' it is a translation of 'celtis.' Moxon names the socket chisel, ripping chisel, paring chisel, skew former, 'mortess' chisel and gouge[2].

The 'polyff' in the *Debate of the Carpenter's Tools* was a pulley, the 'wyndas' a windlass and the 'crow' a lever for lifting the ends of heavy timber. The 'lyne and chalke' says:

> I shall marke welle upone the wode
> And kepe his (the carpenter's) mesures trew and gode.

The above tedious extracts from long-forgotten poems and dictionaries show that the English wright or carpenter of a thousand years ago had axes of several kinds, saws, boring instruments such as augers, some kind of instrument for boring and shaving, and possibly some kind of instrument for grooving. Apart from the difficulty which the compilers had in fitting monk-Latin names to English tools, it is also probable that the tools had not become standardised as to shape, or quite specialised as to use, during the period covered by the vocabularies, as is the case with the thatcher's tools at the present time[3].

As the carpenter's principal tool was the axe, so his principal building timber was the oak, which was the most abundant of British trees on non-calcareous soils, before the destruction of the forests for fuel and for ship building[4].

English oak cannot be thoroughly seasoned, and oak timber which has been fixed in a building for centuries may warp and twist as if it were new, when cut up and used again. So the English carpenters used it when it was cut down, and the durability of their work is the answer to possible modern objections. As an example, the accounts of the churchwardens of Ludlow show that,

---

[1] Mr Robert Phillips in *Builder*, 14th July, 1911.

[2] *Mechanick Exercises*, pp. 72 et seq.

[3] The tools of the Teutonic carpenter before the eleventh century, have been discussed by Moritz Heyne in his *Deutsche Wohnungswesen*, p. 76.

[4] Harrison, the Elizabethan author of a description of England attached to Holinshed's Chronicles says that sallow, willow, plum tree, hard beam and elm were formerly used for ordinary buildings, but now they are rejected, and only 'oke anie whit regarded.' They can hardly be called building timber in the modern sense, although they were used in building.

in the year 1542, the wardens bought a piece of unsquared timber
for a beam, then they paid ten pence 'for the squaryng of hym
in the woode,' and a like amount 'to Rolande Huntt for the
carege home.' Then the beam was fixed in the church and
carved, the latter work having to be done by candle light. The
'squaryng' in the wood would reduce the difficulty and cost of
carriage, matters of importance when roads were bad.

Monsieur C. Enlart[1] says that the carpenters of the Middle
Ages chose the heart of the wood and burned the remainder, and
for that reason their works were very strong, and from it came
about also a progressive disafforestation. Some of the older
buildings of South Yorkshire, however, still retain pieces of the
bark on their timber and therefore must be of a time when the
bark was not of great value. The introduction of barking the
timber for tanners is said to have brought about felling in Spring[2],
and this was recommended by Evelyn in the seventeenth century.
Tusser, in the preceding century, had recommended that wood
should be sawn in September, in order to have unshaken timber.
If kept dry, English oak improves and hardens with age. At the
end of the seventeenth century, Moxon described two methods of
seasoning timber.

The largest logs of British oak were, perhaps, eight obtained in
the year 1322, by Alan de Walsingham, 'after many tedious
personal journeys,' for the angles of the lantern at Ely cathedral.
They are 2 ft. 9 in. × 1 ft. 9 in. × 50 ft. long and sapless[3].

During the seventeenth century oak timber became too expensive
in South Yorkshire to be used in buildings with the old prodigality,
and there was a quite gradual change from the former universal
use of locally grown oak to the present almost universal use of
'Baltic' timber from coniferous trees. In the Middle Ages much
timber was imported from the Baltic and Scandinavia under the
name of 'Estland' or 'Estrich' timber. The Eastland Company,
in the time of Queen Elizabeth, traded with both sides of the Baltic,
and the word Estrich seems to have been a variant of Austria, and
to have meant eastern realm. It can hardly be derived from
'astrich,' a form of floor, in which the materials were arranged in
a pattern and borrowed from the Romans by the Germans. At

[1] *Manuel d'Archéologie Française*, II, p. 185.
[2] *Dictionary of the Architectural Publication Society*, s.v. Seasoning.
[3] Ibid. s.v. Oak.

the Cloth Hall at Ypres, the writer was informed that the great beams of the roof, twelve and a half metres in span, that is about forty-one feet, were from oak trees from Norway, and that the documents relating to the purchase were still in the possession of the municipality[1]. M. Enlart says that mediæval France took iron from Spain, tin from England, and lead from Denmark. Like Ireland and Holland, Denmark also furnished much timber for French buildings: the timber of Denmark was very much used, and he names various works where it was to be found. 'According to Victor Gay it was oak timber with large " mailles," analogous to our oak of " Hollande " and Hainault. It is to be remarked that all these exportations were by water[2].' In the Middle Ages oak trees grew in Denmark and on the shores of the Baltic, where there are now only heaths, and forests of coniferous trees. As our own supply of oak was so excellent and abundant, it is probable that the Baltic timber then imported was generally handsome oak for ornamental purposes, and very rarely coniferous timber. As such it would not be used in the ordinary buildings. Thorold Rogers found that the Hansa towns sent to the great fair of Stourbridge 'the finest kinds of ornamental timber[3].'

It was usual in building to use the same timber over and over again, and with the gradual decrease in the size of scantlings, the use of the old timber compensated somewhat for the declining supply of new. An early mention of fir is in a record of the purchase of two hundred Norway boards of fir ('bordos de Norwagia...sapio') at Southampton in the thirty-seventh year of Henry III, that is the year 1253[4]. In the North of England the change from native oak to foreign coniferous timber for building purposes coincided with the culmination of the Renaissance. Baltic timber was coming into Yorkshire in some quantity in the seventeenth century, for Henry Best, an East Yorkshire farmer who wrote various instructions as to building, in the year 1641[5], and among them advice 'for choosinge and buyinge of firre-deales,' says that 'firre

---

[1] The Cloth Hall has been totally ruined by the German bombardments. The splendid roof was burnt early in the war; it was about six centuries old and the hips must have been as old and as large as any in Europe.

[2] *Manuel d'Archéologie Française*, I, p. 78.

[3] *Six Centuries of Work and Wages*, p. 151.

[4] T. Hudson Turner, *Domestic Architecture in England from the Conquest to the end of the Thirteenth Century*, p. 241.

[5] *Farming Book*, published by the Surtees Society, pp. 110 et seq. and 125 et seq.

deales are accounted better for bordeninge with then oake that
hath not had time for seasoninge, because that when oake cometh
to dry it will shrink, cast, draw a nayle, and rise up at an ende or a
side.'

They go, he says, to Hull 'usually for this commodity' (deals),
'which is brought from Norway. In choosinge of good deales, all
these thinges are to bee considered: that they bee reade' (red) 'deale,
which are allmost as durable as oake, and will not worme eate soe
soone as white deale: besides, they are handsomer and better, both
for smell and colour, and (for the most part) better flowred: that
they bee full twelve foote longe, full twelve ynches in breadth,
somewhat more than ynch thicke: square i.e. as broad att one ende
as the other, and then is there noe waste in them: not shaken,
i.e. cracked and flawed: not knotty: if they bee thus, then the
raffe-merchant may lawfully stile them good deales, and such like
deales can seldome bee bought under £4 10s. or £4 15s. the
hundreth' (or six score). Best gives as one reason for buying
seasoned deals that they are far lighter to carry, and so cheaper.

The Baltic timber came over in an ever-increasing supply,
filling the place of the fast disappearing English oak, and William
Marshall, an agricultural topographer, writing of Yorkshire in the
year 1796, says that 'Floors have been laid with it' (deal) 'for
nearly a century, and of late years it has been used for almost
every purpose of building[1].'

The extension of the building trade, which accompanied the
prosperity of the nineteenth century, caused a great demand for
the cheap and easily accessible timber of the Baltic lands, and
gradually the best timber was used up: and for many years coniferous
timber has been imported from the extensive forests of the American
continent. At first the importation was not made in the place of
the Baltic timber, so much as to obtain showy and ornamental
timber at a cheap price, as in the case of pitch pine. At a time
when everything mediæval was fashionable in English architecture,
this wood, of course unknown before the discovery of America,
curiously enough became that which was most used in all the
cheaper buildings of the Gothic Revival. According to the
Dictionary of the Architectural Publication Society, pitch pine
was almost unknown to the London market in the year 1835
This dictionary quotes Warburton's *History*, published at Dublin

[1] *Rural Economy of Yorkshire*, I, p. 108.

in 1818, which states that pitch pine was preferred then to oak for durability. Time has shown, however, that pitch pine is not thoroughly suitable to structural work as it dries very much and loses strength in so doing.

The conversion and preparation of timber into scantlings in the ports before shipment has been the means of depriving us, practically, of all large Baltic timber, such as the formerly well-known and generally specified timber shipped from Memel, Danzic or Riga, and where comparatively cheap timber of large scantling is required, recourse has been had to the coniferous timber of America, some of which can be obtained in very long lengths[1].

The wood imported from the Baltic is of younger growth, and the desire to maintain the scantling of the timber, with smaller trees, leads to the use of a greater inclusion of sap wood than in the times when the trees were allowed to grow to a larger size.

It has been well said that 'the growing difficulty of obtaining satisfactory material for large panels has led to the use of thin layers of wood, such as birch, consolidated under enormous pressure, and described as 3, 4, or 5 " ply " board. The arrangement of the grain in different directions in different layers adds to their strength and diminishes their tendency to shrink and warp[2].' The craftsmen of the Middle Ages knew this but the modern method of cutting wood for 'ply' work in the roll was beyond the capacity of the mediæval carpenter, who used flat boards.

As we have seen, the supply of English oak decreased with its great use as ship timber for the British fleet during the Napoleonic wars, and afterwards in the extension of the mercantile marine, before the use of iron for ship building. But the demand for oak, for many purposes, continued and the supply was met from the oak forests of East Central Europe, of which the timber is inferior to that of England. The demand has also been met, to a smaller extent, from America.

The most important of the foreign woods in the past was mahogany. Dr Whitaker, in the year 1818, considered that 'the introduction of mahogany about a century ago formed a new aera in the history of internal decoration[3].' It then superseded oak as the fashionable wood in work of a more expensive kind, but it

---

[1] Herbert A. Satchell, *Building Construction*, II, p. 217.

[2] Ibid.

[3] *History of Whalley*, p. 503.

has always been too expensive for use in the ordinary buildings, except in a small quantity as for the handrail to the staircase.

The nineteenth century saw the introduction into England of woods from all parts of the world, but their use is so recent that they can hardly be said to have any history.

CHAPTER IX

WALLS

Wooden walls—Block house construction—Greensted church—Descent of stud
partitions—Posts and panels—Varieties of boarded partitions—Ground
walls or foundations—Stone from the surface of the ground—The use of
clay, earth, and mud for mortar—Influence of freestone walls.

In Anglo-Saxon and semi-Saxon glossaries from the ninth to
the twelfth centuries the Latin word 'edificium,' which means 'a
building,' is translated by 'getimbrung,' that is, literally, 'timbering.'
The word does not imply that all buildings, at that early time, were
of timber : it was originally applied to buildings of any materials,
but in course of time the word, and its allies in other Teutonic
languages, were used for building in wood, because 'the term for
construction became identified with the material most generally
employed.'

During a long period wood was our national building material,
not only in Great Britain but in Ireland also. There are instances
of the old Irish preference for wood in early documents. Thus
according to the Book of Armagh, when St Patrick went to
Foirrgea he made there a squared earthen church of soil because
there was no wood near[1]. The Irish preference for wooden
buildings would necessarily involve the use of plates or sills of
wood, which is not available in shapes suitable for apses, and it is
therefore possible that the square east end of Celtic churches was
primarily due to the exigencies of building construction, and at
a Romano-British house in Bucklersbury, London, all the walls
had been of timberwork, with the exception of the apse[2].

In France timber was in general use for building in early times
and M. Enlart says that 'a certain number of Merovingian churches

[1] Quoted in *Notes and Queries*, 9th Series, II, p. 410.
[2] John Ward, *Romano-British Buildings and Earthworks*, p. 250.

were constructed of wood, including the cathedrals of Rheims and Strasburg, and the first church of Tongres, constructed of well planed planks ("de tabulis lignis levigatisque")....An indirect proof of the frequency of buildings of wood resides perhaps in the application to the sculptures of stone from the fifth and sixth century to the beginning of the twelfth century, of a technique which seems made for sculpture on wood. Unhappily, this category of buildings can only be pointed out, as all attempts at the restoration of their forms would be pure conjecture[1].'

The oldest building with wooden walls in England is the little church of Greensted, near Ongar, in Essex. It has been cited as an English example of the so-called 'block house' construction[2]; but in true block house construction—the 'block bau' of the Germans—the timbers are laid horizontally, and at Greensted, on the contrary, they are upright.

The true 'block house' construction is of great antiquity. Vitruvius, the Roman writer on architecture, described it in its simplest form as used by the Colchi. He said 'The woods furnish such abundance of timber that they are built in the following manner. Two trees are laid level on the earth, right and left, at such distance from each other as will suit the length of the trees which are to cross and connect them. On the extreme ends of these two trees are laid two other trees transversely: the space which the house will enclose is thus marked out. The four sides being then set out, towers are raised whose walls consist of trees laid horizontally, but kept perpendicularly over each other, the alternate layers yoking the angles. The level interstices, which the thickness of the trees alternately leaves, are filled in with chips and mud[3].'

In later developments of the block house construction, the level interstices were done away with by cutting the ends of the logs in various ways to make joints, thus bringing the logs close together, as in Figs. 61 and 62, and, further, block house and post and truss constructions were combined by filling the panels, between the frames, with horizontal beams.

The block house type of construction spread in Europe, westwards to Switzerland, and northwards to Scandinavia. The

---

[1] *Manuel d'Archéologie Française*, 1, p. 125.

[2] G. Baldwin Brown, *Arts in Early England*, II, p. 41.

[3] Bk II, chap. I.

European settlers took it with them across the Atlantic to the new and well-timbered continent of America, and the Russians have carried it across Asia, as far as the Pacific Ocean. It has therefore girdled the world, and in doing so it has followed the forest belt of coniferous trees of the North Temperate Zone, being obviously a form of construction which is only possible where there is an abundant supply of timber.

There is no satisfactory evidence that this form of building was ever in use in England in any of its forms, although fine timber was formerly so abundant. The place-name Bullhouse has been supposed to be evidence for block house construction, because, in Scandinavia, a block house is called 'bul-hus,' that is, 'bole house': but it has been shown[1] that 'bul-hus' referred originally to any building of wooden boles, and the place-name Bullhouse, if it is connected with building at all, refers more probably to a construction of the Greensted type. Mr Frederic Seebohm[2] and Dr C. M. Andrews[3] considered that the word 'stoc-life' in King Alfred's *Blossom Gatherings from St Augustine* meant log-house: but here, again, if a building is referred to at all, stake-house is a more reasonable interpretation.

The church at Greensted, from which we have rambled, has its walls formed of half trunks of trees set upright, with the split faces inwards. The upper end of each half trunk originally was roughly tenoned into a plate and the lower end into a sill. There were two pegs to each tenon and both the plate and the sill were tenoned. The sill was laid upon the ground, but in the course of time it rotted, as did the lower portions of the trunks, and in one of the various 'restorations' which the church underwent in the nineteenth century, the sill and the rotted parts of the trunks were removed, and a low dwarf wall built, on which a new sill was placed. The roof was also renewed at one of the 'restorations.' Each half trunk is connected to its neighbour by a grooved and tongued joint, and at the angles of the building three quarter trunks are used in order to keep the continuity of the walls, inside and out. It is now claimed that the church was put up as a temporary resting place for the corpse of St Edmund, when it was removed in the year 1013, but the time and labour required for the erection of such

[1] K. Rhamm, *Urzeitliche Bauernhöfe*, p. 516.
[2] *English Village Community*, 4th edition, p. 170.
[3] *Old English Manor*, p. 109.

a building in early times make the theory of a temporary building
untenable. The chancel is a later addition, and the original part
is an oblong and rectangular apartment, which is as much like a
strong hall as a church. There is, indeed, no evidence by which
to date the building and it may have been built at any time before
the coming of Gothic, as so little is definitely known of the more
expensive methods of timber construction in this country in early
times. However, in favour of its early date is the fact that Moritz
Heyne believes that Hart Hall, in the early poem *Beowulf*, was
built of upright trees, as at Greensted[1]; he also believes that the
timbers were fastened together with iron cramps. Heyne says
that in Germany the ends of the trees were set in the ground,
their ends having been previously charred to prevent them from
rotting[2].

The late Mr Ernest Godman[3] also did not agree with the
identification of Greensted church as the chapel for the body of
St Edmund, and he remarked upon its dedication to St Andrew.
He found that the timbers of the walls had been dressed with an
adze internally, after they had been split.

It has been pointed out 'that the same compact system is
shown in the representations on the Bayeux Tapestry of the
timber structures that surmount the moated mounds, several of
which are figured in the needle work. They seem put together
just in the manner illustrated at Greensted[4].'

The post of the wall against which was hung the miracle-
working earth from St Oswald's grave, as described by Bede, was
probably similar to those at Greensted[5]. It is 'posta parietis,'
that is wooden wall-post, in the original version, and in the Anglo-
Saxon translation it is 'studu thaes wages,' that is, stud of the
wall. The originals of this type of wall are simply fences composed of trunks of trees set upright, close together, and in a line.
The Greensted church wall is similar to the stockaded ramparts
of the English boroughs and the Norman *mottes*, from which the
walls of their timber superstructures were copied. M. Camille
Enlart has quoted from a biography of Jean de Warneton, bishop
of Térouane, of about 1100 to 1130 A.D, an example of a wall of

---

[1] *Lage der Halle Heorot*, p. 32.   [2] *Deutsche Wohnungswesen*, p. 18.
[3] *Essay on the Characters of Mediæval Architecture in Essex*, p. 30.
[4] Professor Baldwin Brown, *Arts in Early England*, II, pp. 41, et seq.
[5] *Ecclesiastical History of the English People*, Bk III, chap. X.

wooden boards very strongly compacted together ('vallo ex ligneis tabulatis firmissime compacto')[1]. When such timbers are arranged so as to form an enclosure, and roofed over, they form a strong and simple but costly building.

We shall see that one line of descent of the later mediæval 'half timber' work is from such timber walls as those at Greensted. It has already been shown that the boughs used as rafters were squared in course of time, and the same development took place with the walls: from the whole tree trunks of the early Germanic walls descended the half trunks of Greensted church, with results which were economically and æsthetically favourable, and in a further development the half trunks were squared. Mr S. O. Addy, in his *Church and Manor*[2] says that either in the reign of Alfred or of Canute, it is not clear in which, a church was built at Bury St Edmunds, then called Bedricsworth, of 'wooden boards.' On a certain occasion it was attacked by robbers, one of whom tried to dig under the wall with spades and mattocks. The noise which the robbers made awoke a man who was sleeping in the building. In buildings upon which less labour was expended than upon the church, tree trunks like those at Greensted and such boards as those at Bury St Edmund's, would not be jointed, but merely caulked with some sort of adhesive material, such as clay or earth or moss, as is done still in the wooden buildings of Canada and the United States. In the Preceptory at Chilborn, Northumberland, a building of the end of the fourteenth century, 'the partitions were of oak plank placed in a groove at top and bottom, with a narrow reed ornament on their faces, 3 in. in thickness, placed at a distance of 12 in. and the interstices filled with loam[3].' These Chilborn partitions appear to have been inside the building, and old methods of forming outer walls are used in the interiors of buildings long after they have been banished from the outer walls in favour of improved and stronger methods. Buildings of this kind with outer walls formed of upright studs, filled in between with mud, are still to be seen in Denmark.

The stud partitions which were so usual for the interior divisions of English buildings, until the comparatively recent introduction of patented partitions, are the descendants of solid walls of wood like those of Greensted church. It is well known that the older the

---

[1] *Manuel d'Archéologie Française*, II, p. 499.　　　　[2] p. 44.
[3] *Dictionary of the Architectural Publication Society, s.v.* Partition.

studwork, the closer are the studs together, the larger are their scantlings, the more do upright timbers predominate, and the less are they inclined or horizontal or patterned. The studs in old buildings are usually pegged to the plates or sills with two pegs, the number employed at Greensted church.

An important difference arose in the course of the development of the 'half timber' construction from the walls composed of tree trunks: in the latter the trunks form the walls, in the former the studs are only the filling-in of a framework, which was the result of that continuous trend in the direction of economy in the use of materials, combined with specialisation of function, which marks the history, not of building only, but of the other *terrains* of human culture.

Another development arose, due to the specialisation of function. At Greensted church the posts are all alike, but in the buildings on crucks, and also in the post and truss buildings, the uprights are graded according to the work which they have to do. The crucks or posts which carry the roof are the largest, then next in size are the posts which carry the end of the tie beam in the cruck building, and, smallest of all, are the studs of the outer wall which carry the boarding or the plaster.

A greater use was made of horizontal timbers in the sixteenth century and these, when combined with the upright studs, formed panels which were generally filled with plaster. In Lancashire the construction was known as 'post and petrel' or 'post and petrail,' from the French word 'poutreille' a cross beam. Such walls are also known as 'post and pan,' the latter word being said to mean pane or panel. In South West Surrey the panels became larger in the sixteenth century, and the spaces were decoratively filled with curved pieces of timber, using up 'crooked bows and limbs of trees that would have been useless for ordinary buildings[1].' The greater cost of timber was responsible for this and for the increased width of the panels.

An instance of that retreat of older forms to the North and West of England, which has already been mentioned, is seen in the half timber buildings of the Western Midlands, of the fifteenth and sixteenth centuries, where the timbers are larger and more massive, the details of the moulded parts coarser and less elaborate,

---

[1] Ralph Nevill, *Old Cottage and Domestic Architecture in South-West Surrey*, p. 19.

and the carving, as a rule, more primitive than in the South Eastern counties[1].

When the bearing of the beam was over twenty feet between the principal posts in the post and truss buildings, smaller posts called prick posts were sometimes inserted between the large posts. Moxon advised that these intermediate posts should not run up the whole height of the building as did the corner posts, but only from storey to storey. The studs were also called quarters and puncheons, and in Moxon's time single quarters were 2 in. × 4 in. and double quarters were 4 in. × 4 in. in scantling.

We have seen that the wall studs of the 'stridsuler' buildings of Denmark were the last part of the building to be erected[2]. In South Yorkshire the writer has found that, wherever there is a building 'on crucks,' it is accompanied by the tradition among the old people that the walls were put up after the roof, and in cruck buildings as they exist to-day, this is literally the case, for the walls seem to be always of comparatively modern stone work or brick work, but the tradition is much older than the modern walls. Even in a few examples, where the stone walls are possibly of the date of the crucks, they have evidently been fitted to the timbers and, we have already seen[3] that in the large barn at Peterborough, built in the year 1307, the wooden posts and roof trusses were independent of the stone walls, and must have been erected before them, as, at the destruction of the building, the wooden posts and roofs still stood securely when the stone walls were pulled down.

Byrhtferth, an Anglo-Saxon author of the tenth century, whose description of the building of a house has been printed in the German periodical *Anglia*, VIII, says, 'The rafters are fastened to the ridge tree and propped underneath with cantles,' (which were studs or quarters), 'and then the house is agreeably adorned[4].' This shows that the method of inserting the wall studs, after the roof was reared, is at least a thousand years old in this country, and its existence in Denmark gives it a wider significance. In ancient times the roof was regarded as the most important part of the building, whereas, at the present time, the building laws are more concerned with the walls. The explanation lies in the gradual evolution of modern buildings from those early structures

---

[1] J. Parkinson and E. A. Ould, *Old Cottages, Farm-Houses and other half-timber buildings in Shropshire, Herefordshire and Cheshire*, p. 2.
[2] Ante, p. 21.   [3] Ante, p. 74.   [4] p. 324.

which were all roof and had no walls, as we understand them, of which the development has been traced in the preceding chapters.

The 'cantles' of Byrhtferth's house lead us naturally to another descendant of the Greensted church type of wall, the boarded partition.

It has already been shown that the word 'tabulata' was applied in early times to the walls of trunks of trees and in the fifteenth century the word was translated 'burd-wogh,' that is, a board wall, or, as we should now call it, a boarded partition. The development of the mediæval boarded partitions follows the general rule as to a

Fig. 43. Houses at Wigtwizzle, South Yorkshire. The house on the right contains an ancient wooden partition.

progressive economy, and they became slighter in the course of time. In the simplest type all the upright pieces, the boards, are of the same thickness. Inside the house shown in Fig. 43 there is a boarded partition, of vertical oak boards, alternately grooved and tongued. Each board is of the same thickness, and as the edges have not been squared, the grooves and tongues follow the wavy edge. It is possibly a survival from an earlier building on the same site. The dull appearance of this early plain boarding was some-times relieved by chamfering half the width of each board in the same direction on every board. Another variety of plain boarding is the so-called 'in-and-out' boarding, in which the boards are not placed in line, but alternatively backward and forward, or 'in-and-out.' This produces light and shade and gives a spurious appearance of thickness. An example of this, exposed to the weather and falling to decay, was seen by the writer in the ruined portion of Plas Twdr, Henryd, near Conway, on 11th Oct. 1913.

In another form of boarding some degree of light and shade was obtained without the labour of chamfering by setting each board slightly out of line: this may be seen in old buildings in

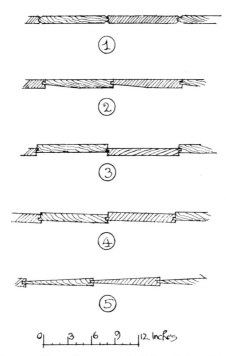

Fig. 44. Plans of varieties of old boarded partitions.

South Yorkshire, and in a closely allied variety, also used in South Yorkshire, each board of the partition tapers in thickness in the same direction and so light and shade is obtained in another manner. Very similar boarding may be seen in the so-called house of Memling, at Bruges. These varieties of boarded partitions are shown in plan in Fig. 44, and probably many other varieties are still to be found in old buildings in the more remote nooks and corners of England. They show the inventiveness of the old carpenters within a very limited range.

A portion of the gable which remains at Greensted church has been thought to be original. It has been described as formed of two vertical layers of planks fastened together with treenails, and as the planks were not long enough to reach the full height, they were arranged to 'break joint' both perpendicularly and horizontally[1]. The shortness of the boards is probably due to them having been split or rived.

The boarded partitions followed the stud work in their development from partitions formed entirely of upright pieces to panellings with styles and rails, and even with curved pieces, analogous to the shaped pieces of the walls and gables of the later timber houses. The progress may be traced step by step from the simple 'burdwogh' of timber, like railway sleepers set on their ends, of pre-Conquest times, to the elaborate and heavily moulded panelling of the English Renaissance.

The former remoteness of the country about Sheffield is responsible for the survival, to the present day, of curious examples of building construction. Thus, at a house in the Little Don valley there is a partition of oak with heavily moulded panels, running the whole height of the partition, and without cross rails. Such a partition is a result of the application of the mouldings of the Renaissance to the early method of fitting an upright board between the studs or quarters of a wall.

A form of partition has been described which was less elaborate than that of the Little Don valley panelling. It was 'of oak, very roughly made, nearly alike on both sides, formed of boards 14 in. wide, fixed in vertical grooves in stout uprights, which are 6 in. wide, with chamfered edges, having triangular, or sometimes leaf-shaped chamfer stops about 7 in. from the bottom, the whole fixed by means of mortices in a horizontal beam resting on the floor, and above in a horizontal beam chamfered over the spaces between the uprights, with short returns to meet the chamfered edges of the uprights[2].' This is a mediæval ancestor of the partition of the Little Don valley, in which the mouldings of the Renaissance have not yet supplanted the chamfers of the Middle Ages.

Allied to the combination of framing with horizontal logs, already described, was the form of wall in which timber framing is covered with horizontal boards, and which is still very common in

---

[1] E. Godman, *Essay on the Characteristics of Mediæval Architecture in Essex*, p. 30.
[2] *Notes and Queries*, 9th Series, III, p. 265.

the Southern counties of England.   Probably the earliest example
ever recorded was one found in Drumkelin Bog, Donegal, Ireland,
in the year 1833[1].   At the time of its discovery it was believed
to have been constructed in the Stone Age.   The walls were
formed of horizontal planks secured to rebates in the framing.
At the Lawsing Stedes barn at Whalley, Whitaker stated that
' here, instead of walls, there are nothing but oak boards, fixed
diagonally, like a Venetian blind[2].'   In one part of a building
' on crucks ' in Ewden, South Yorkshire, such boarding as that

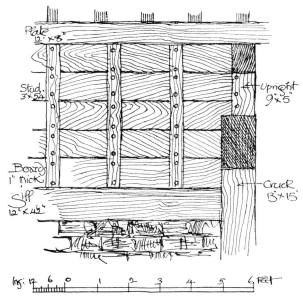

Fig. 45.   Elevation of an ancient external wall or partition of horizontal
boards, at Renold House, in Ewden, South Yorkshire.

at Whalley still remains, but it can only be seen from the interior
as a wall has been built on its outer side.   Each board is pegged
by two pegs to every stud (see Fig. 45), and although the
wall protects the oak boards from the weather, they are worm-
eaten and decayed, and evidently of great age.   This is the only
example known to the writer in the Sheffield district, but in other
ancient timber buildings in the neighbourhood the writer has found
pairs of peg holes in vertical rows on the upright timbers, which at

[1] *Archæologia*, XXVI, pp. 361 to 367.
[2] *History of Whalley*, published in 1818, p. 500.

present serve no useful purpose, so that such a form of boarded wall appears to have once been usual; when the boards decayed they were replaced by the present stone walls, which had come into general use in South Yorkshire by the beginning of the eighteenth century. It is curious that the horizontal boarding should have survived in the South rather than in the North, but this may be a result of the milder climate.

Stone was only used for the foundations, in ordinary English buildings, up to the time of the Renaissance, and walls of worked stone, or 'ashlar,' are scarcely to be found in the buildings of older England, except in those of importance. The earliest use of stone seems to have been to form a foundation on which rested the sill of the wooden framework of the wall : such stones were used without having been dressed, and were gathered from the surface of the land, from river beds or from boulder clay. Stones lying on the surface of the land are known in South Yorkshire as 'day stones,' because they lie open to the air and have not to be quarried, and also as 'groundfasts,' because they are sunk partially in the ground.

In many of the old stone cottages of South Carnarvonshire, enormous stones have been used for the foundations, and these generally extend up to the ground floor level, above which is an ordinary wall of smaller stones. The use of these large stones is a survival from the times when the walls were built of wood and wattling, such as now only survives in external use in the gables, as in a house near the church at Llanelian, near Colwyn Bay. In a barn 'on crucks' at Hall Broom, in South Yorkshire, a foundation course of such large rough stones still remains, and such a stone foundation was the 'ground wall' of the Anglo-Saxons and of their successors in the Middle Ages. In the Ripon accounts John Mason is paid for making 'divers walls, which are called Grund-walles,' in the year 1399–1400[1]. Ground walls, without mortar, of quadrangular houses of post-Roman date have been found in the English Lake district[2].

Such rough stone foundations were liable to settlements, and in proofs of the ages of heirs for Northumberland of 8 Henry IV, it is stated that 'Henry Chester underpinned anew a house in

[1] *Memorials of Ripon*, published by the Surtees Society, III, p. 132.
[2] By Mr H. S. Cowper, F.S.A., who has described them in *The Cumberland and Westmorland Antiquarian and Archæological Society's Transactions*, XVI (1901).

Chevelyngham : he was chamberlain the same day[1].' This means
that he drove thin pegs or wedges of wood between the wooden sill
and the stone foundations, and in this we have the origin of our
modern 'underpinning.' What is now work requiring the greatest
care and skill was in the fifteenth century the casual labour of a
' chamberlain.'

In the year 1796, H. Kent, an agricultural writer, advised that
all the ground sills or foundations should be kept constantly tight
to prevent the wall or upper part of the building from getting out
of the perpendicular or warping[2]. The superstructure was, of
course, a wooden framework. The sills are named 'Ground-
Plates' by Moxon, who says that they are made of the same
scantlings as the principal corner posts, morticed together at their
angles, and pegged with 1½ in. pegs.

There are probably few localities where the use of stone as the
principal material for the walls of minor buildings is of any
antiquity. In West and South Yorkshire, well endowed by Nature
with good building stone, the walls of the pre-Renaissance buildings
were generally formed of wood or wattle and in South Pembrokeshire, forty years ago, the cottages were built of clom, that is, clay
mixed with chopped straw, 'which is strange, seeing that in many
neighbourhoods stones are apparently more plentiful than soil : but
clom being the cheaper material to work, of that the peasants'
dwelling was generally constructed[3].' The walls of some parsonages
in the stony county of Cumberland were made of clay in the
seventeenth century[4]. Cornwall is another stony county where
clom was used equally with the rough granites of the district until
quite recently. 'The humble Cornish builder of ancient and
modern times set in huge masses of granite just as he found
them, and the larger they were the better they answered his
purpose : if he could make three or four great blocks of stone
form a wall, the less labour and skill was required in building,
and the main object was obtained[5].' This agrees with the rule that
the fewer the joints the more primitive is the construction, and is

[1] *Archæologia Aeliana*, XXII (1900).

[2] *General View of the Agriculture of the County of Norfolk*, p. 114.

[3] E. Laws, *Little England beyond Wales*, p. 403.

[4] Bishop Nicolson's *Miscellany Accounts for the Diocese of Carlisle*. Ed. R. S.
Ferguson, pp. 16 and 18. The parsonages were those of Burgh-by-Sands and Kirk-
Andrews-on-Eden.

[5] J. T. Blight, *Churches of West Cornwall*, p. 166.

not entirely due to a wish to save labour although methods of building are always conditioned to some extent by the materials. According to Mr Blight, it is difficult to tell the age of a modern Cornish cottage if in ruins, because the same kind of stone wall has lasted from the time of the hill forts and hut circles up to the present.

The walls of large, undressed stones are historically ancient, as according to Bede (*Acta Sanctorum*), the walls of the monastic establishment, founded at Farne in the year 684 by St Cuthbert, were formed of rough stones and earth, and some of the stones were so large that four men could scarcely lift them.

Many early stone walls were plastered; a very fine plaster was used in Anglo-Saxon times on stone walls and according to Mr H. W. Brewer, rubble walling in Germany, during the mediæval period was always covered with plaster both inside and outside. It is a matter for regret that so much old plaster was stripped from the walls of English churches during the 'restorations' of the nineteenth century.

The stones for the earlier walls were obtained in the same manner as the stones for the foundations or ground walls and the method is still used in out-of-the-way places. In South Carnarvonshire the writer has seen masons breaking up the glacial boulders lying on the surface of the land, in order to use them in the walls of a farm house, which they were about to build on the site. The masons were both quarrymen and wallers and the two trades had not become separated. In Lewis there is a much more primitive stage in the evolution of industry, for the huts, or 'clachans,' are built by the peasants themselves, there being no craftsmen who devote their whole time to building.

The stones of the early walls were put together without mortar as are the walls of the Lewis 'clachans,' and the field walls of the Sheffield district at the present day. Old methods of walling sometimes survive in the fences after they have become obsolete in the buildings, and so they may give a clue to the earlier methods of building in the district. In Leicestershire in the year 1809, Wm Pitt found that the cottages often consisted of mud walls and thatch, and that the fence walls in the South and East of the county were also of mud, coped with clods or thatch, but such buildings were being gradually removed[1].

[1] *General View of the Agriculture of the County of Leicester*, p. 26.

The rough stones of the walls of a primitive kind do not always fit accurately together, and the stones were bedded in earth at a very early period. W. Marshall, writing of the *Rural Economy of Yorkshire* in the year 1796, says that 'formerly ordinary stone buildings were carried up entirely with "mortar," that is common earth beaten up with water, without the smallest admixture of lime. The stones themselves were depended upon as the bond of union, the use of the mortar being merely that of giving warmth to the building and a degree of stiffness to the wall[1].' Here there was only the idea of bedding and not of adherence, but the old builders early saw the advantage of stickiness in mortar, and so used clay, mud, lime and cow dung; they seem to have thought only of the present, knowing nothing of the chemistry of mortars, and making little distinction between lime and other materials. This is the reason for the apparently rubbishy mortar in some Anglo-Norman great stone piers and walls rather than a desire to 'scamp' the construction. A description of the making of Yorkshire 'mortar' in the year 1641 has been left by Henry Best, the East Riding farmer[2], whose instructions for buying deals have already been given. Two 'coupes' were to be sent into the fields and loaded with 'clots': the loads were to be thrown down near to water, where three men with clotting mells broke up the clots, after which they were watered and tewed and left to be tempered. Such was the procedure in summer; but in winter the material for the mortar was to be obtained by shovelling up dirt free from gravel and stones, or by digging down an old wall, or by graving up some earth. The watering was to be done over night, and the mortar allowed to steep all night: it was to be made so soft as almost to run, then the water was to be allowed to 'sattle.' Best says that mortar makers always had one or two old spades, a little two gallon 'skeele' for water, and old scuttles to carry the mortar. It was usual to put a handful of dry straw into the bottoms of the scuttles.

Mr John Ward suggests that in some Roman mortar of earth and lime, 'the latter has sometimes been removed by the solvent action of the moisture of the soil, leaving an earthy residue which has misled observers into thinking that puddled earth was used for mortar[3].'

[1] I, p. 101.
[2] *Farming Book*, published by the Surtees Society, p. 145.
[3] *Romano-British Buildings and Earthworks*, p. 256.

It is interesting to note that at the Romano-British station of Caersws, the mortar of the walls was everywhere of clay, except in the strong room, where lime mortar was used[1]. There is an old Welsh legend which explains the discovery of mortar in the same accidental manner as Charles Lamb's humourous account of the discovery of roast pork, and the older story of the discovery of glass.

The Latin word 'basis' is 'syl' and 'syll,' that is 'sill' in Anglo-Saxon glossaries, and at the same period 'fundamen' and 'fundamentum' are translated as 'grundweal,' that is 'ground wall' and as 'stathol,' which survives in many dialects as 'staddle,' the foundation for ricks and stacks. A staddle is now usually an arrangement to raise the stack above the ground, and in the records of Queenborough mention is made in the time of Queen Elizabeth of the six and twenty houses, which are commonly called the old staddles ('de viginti et sex domibus qui vocantur the old staddeles, or six and twenty houses'). Probably these were houses raised on posts somewhat after the manner of lake dwellings. The mode has descended in the social scale, as usual, for Moritz Heyne mentions that a palace of Charlemagne was raised on posts in the eighth century[2].

Buildings of which the walls were altogether of stone provided other methods of carrying the ridge-tree than those which have been already discussed : the ridge-tree might be carried by the stone gable walls, or if the walls were not carried up the full height of the gable, the ends of the ridge-tree might be propped by upright pieces of wood standing on the end walls. The former is more usual in England, and the latter in Norway and Sweden.

With the extension of the use of stone for walls there came experience as to the best method of putting it together to form stable and weather-proof buildings. In every district which produces building stone there is a local variety of walling which is conditioned by the kind of stone found in that district, and it is a matter for regret, from the æsthetic standpoint, that this is being broken down in our day by the standardisation which is produced by text books and examinations, and by that ease of communication between districts which facilitates the transport of alien materials and methods. In South Yorkshire stone walls began to be generally

---

[1] Prof. Bosanquet at British Association Meeting, 1910.
[2] *Deutsche Wohnungswesen*, p. 16.

used for ordinary buildings in the seventeenth century, and they were constructed of stones from the easily quarried upper beds, which gave walls of good texture and with very fine and level joints. The walls were often left dry on the outside, and the interior was filled in some cases with sand and in others with clay.

The builders in rough stone had the admired works of the freestone masons as patterns, and so there arose forms of walling which are intermediate between rough rubble and ashlar. An

Fig. 46.   Stone houses at Wigtwizzle, South Yorkshire, with stone walling intermediate between rubble and ashlar (see also Fig. 43).

example from Wigtwizzle on the edge of the South Yorkshire moorland is shown in Fig. 46.   The rise in the price of timber in the seventeenth century, which accompanied the destruction of the forests, made it as cheap to build houses of stone as of wood and plaster; nevertheless the adoption of stone was only gradual. In South Yorkshire at first, the ground floor walls only were filled in with stone; later all were constructed of stone except the gables; and finally, as we have seen, all the outer walls were constructed of stone.   In South-West Surrey, according to Mr Ralph Nevill, the

custom of building timber framed cottages certainly continued as
late as the year 1700[1], and in the Western Midlands it lingered
on until well into the eighteenth century[2].

[1] R. Nevill, *Old Cottage and Domestic Architecture in South-West Surrey*, p. 6.
[2] J. Parkinson and E. A. Ould, *Old Cottages, Farm-Houses and other half-timber buildings in Shropshire, Herefordshire and Cheshire*, p. 3.

CHAPTER X

WALLS (*Continued*)

Wattlework and its varied uses—Daubing—Hurdles—Frailty of early ordinary buildings—Varieties of wattled partitions—Their names in the dialects—Walls of earth and mud—Cob buildings and their construction—Lathed partitions—Absence of lime—Plastered panels—Brickwork—The Great Fire of London—Bonds and bond timbers—Size of bricks—Sun-dried bricks in England—Walls of turves.

It has already been shown that some Nottingham masons, in the year 1511, were protected when at work by 'flakes,' which were screens made by weaving twigs in and out of uprights fixed at distances apart. According to the *Oxford English Dictionary*, the senses of the word, of which the variants are widespread in the Teutonic languages, seem to point to an origin in some word meaning to plait. Work of this kind is now generally called wattling when on a large scale, and basket work when small. It is of great antiquity and its use is widespread all over the world for many purposes. Wattles, in addition to their use for walls, have been used in soft ground for a foundation for building and to form weirs when the waters of a river were to be dammed up. An old semicircular vault in the crypt of the deanery at Kilkenny is said to have been turned on a centering of wattlework according to' the *Dictionary of the Architectural Publication Society*[1]; in the Middle Ages, the platforms of building scaffolding were made of wattlework instead of planks, and such platforms were used by country builders as lately as the middle of the last century. Alderman Sir W. Staines is said to have contrived a scaffold of wickerwork to the spire of St Bride's church, Fleet Street, which had been damaged by lightning in the year 1764[2]. This was improved upon by T. Birch in repairing the steeple at St Alban's church : he perfected it in

---

[1] *s.v.* Wattle.   [2] Ibid. *s.v.* Scaffold.

1776, when, for the sum of £20, he erected such a scaffold around the spire of Islington church, within which was a spiral staircase from the tower to the vane[1]

The walls made of basket work or wattlework had been improved, in prehistoric times, by covering them with earth, clay or some other adhesive material. This is now generally called 'daubing,' and, like the wattling, it has a variety of names in the English dialects.

Mr G. W. Smith has described[2] wattle-and-daub discovered at the Queen's Head Inn, Down, Kent, which is said to have been built in the fifteenth century. The rods were 1 in. or $\frac{7}{8}$ in. thick, mostly hazel sticks with the bark on, and both bark and stick were found to be as sound and tough as when it was fixed, and every little vein in the bark perfect. Mr Smith says that they came across several sticks of ash, quite worm-eaten and rotted, and often the only material left in the clay to show their position. Straw was mixed with the pug or clay in order to toughen it, and it is said that the working of the 'dab' into the 'wattle' was done by two men, one on each side, so that they met together on the 'wattle,' and then kept throwing it until they got on the thickness wanted. The front or outside was then combed or lined out with a pointed stick.

Mr Bankart thinks that few specimens of wattlework are now remaining, but it is of course hidden behind the plaster, and so is not apparent. It is still common in many districts of England and Wales. As an example, Rev. R. K. Bolton has described a mediæval cottage at Ffenny Bentley, Derbyshire, in which some of the interior walls were composed of wattle like basket work, with the uprights about a foot apart and about 1 in. thick, the interlacings slightly lighter, composed of unpeeled twigs of what appeared to be hazel[3]. These examples from Down and Ffenny Bentley are of the most usual kind of English wattle and daub.

Wattled walls were used by the Romans, and their liability to decay and act like torches in a fire caused Vitruvius to regret that they had ever been invented. One cause of the decay, he found, was that the walls were always full of cracks. These arose from the crossing of the laths or twigs, for when the plastering was laid

---

[1] *Dictionary of the Architectural Publication Society*, *s.v.* Scaffold.

[2] In Mr Bankart's *Art of the Plasterer*, pp. 55 et seq. There is an illustration from a photograph.

[3] A church in the Peak of Derbyshire in the *Reliquary*, N.S., VI (1900), p. 93.

on wet it swelled the wood, and as the work dried it contracted again, and broke the plastering in doing so[1].

In England the Romans used wattlework for various purposes, as for the lining of wells, as at Caersws. They used it for the walls of buildings, and some of the earliest wattlework remaining in England, was discovered by General Pitt-Rivers at the Romano-British village of Woodcuts Common[2]. He found that the stakes were 0·9 in. in diameter, fixed at 5 in. centres, and the wattles were 0·4 in. to 0·6 in. in diameter, and half the thickness of the daub was 1½ in., giving 3 in. as the total thickness of the partition or wall. At Mount Caborn the stakes were ¾ in. in diameter and fixed at 5 in. centres as at Woodcuts Common : the wattles were one third of an inch in diameter and half the thickness of the daub was 2 in., giving 4 in. as the total thickness of the partition or wall. In these Romano-British examples the wattling does not appear to have been in frames, as the rounded daubing of the ends of the wattling was found. Apparently the ends of the wattle stakes were merely driven into the ground without a sill, as is still the practice of the Albanians, who are the most primitive in culture of all true Europeans. Probably the ends of the Romano-British wattles were formed in the same manner as are hurdles in West Surrey to-day. The hurdle maker uses a 'hurdle frame,' a long shaped block slightly curved to hold the feet of the uprights, which are round rods ; 'the man then weaves in horizontally the smaller split rods till he has filled up the hurdle. When he comes to either end he gives the rod a clever twist that opens the fibres and gives it something the character of a rope, so that it passes tough-stranded and unbroken round the end uprights[3].'

When the twigs or wattles had been interwoven with the posts the hurdle became one homogeneous whole. An early example of this was excavated at the Glastonbury lake village in the year 1896. Amongst the wood and débris underlying the clay of a dwelling mound three hurdles were uncovered : the more complete one measured 6 ft. 3 in. high by 10 ft. 6 in. wide, with an average space between the upright posts of 5 in.[4]

Although many of the more important buildings of the

---

[1] Bk II, chap. VIII.   [2] *Excavations in Cranborne Chase*, I, p. 147.

[3] G. Jekyll, *In Old West Surrey*, pp. 198 and 199. A form of wattling known as wrestling ('rozzling, rostling, rahstling') is also described.

[4] S. O. Addy, *Evolution of the English House*, p. 6.

Romanesque and Early Gothic periods still remain, there is not a single example remaining of the contemporary ordinary buildings. They must therefore have been too poorly constructed to have survived, and if they were merely of daubed flakes their disappearance is easily accounted for.   Such a mode of construction must have been very weak, and the frailty of the early ordinary buildings is apparent from the documents.   The ordinary houses of London in the year 1212 were so unsubstantial that the aldermen of the City were provided with a crook and cord to pull them down when they caught fire, or even if they did not comply with the legal requirements as to fire resisting construction, the mediæval equivalents of our modern bye-laws[1].   In a proof of the ages of heirs of Northumberland of 10 Henry VI, it is stated that so large and strong a wind arose at Morpeth that all the men and women of the said vill greatly feared for the shaking of their frail houses[2], and Hunter noticed a tradition of a village near Conisborough having once been blown entirely away[3].   Such buildings cannot have been 'on crucks,' for as a Derbyshire farmer said to the writer, a traction engine could be run over *them* without hurting them.   The crucks were in regular use for ordinary buildings at the end of the Middle Ages, and in out-of-the-way places they were used until the seventeenth century. Plainly, the ordinary buildings were very weakly constructed before the general adoption of crucks, and no doubt the older and slighter methods lingered, in their turn, in the more remote districts. Froissart tells how, in the fourteenth century, while the French and Scots invaded England, Scotland was invaded by an English army which burnt and destroyed the houses.   On the return of the French and Scots they found that the whole of the Lowlands of that country had been ruined ; but the people generally made light of it, saying that 'with six or eight stakes they would soon have new houses.'   So King Alfred, in his *Blossom Gatherings from St Augustine*, says that of fair rods ('feyrum yerdum') will a man build his village, and in the *Catholicon Anglicum*, some five hundred years later, 'to make a howse' is translated 'palare,'

[1] T. Hudson Turner, *Domestic Architecture in England from the Conquest to the end of the Thirteenth Century*, p. 282.   In St Bene't's Church, Cambridge, a similar hook remains.   It is of iron $1\frac{1}{2}$ in. × $1\frac{1}{2}$ in. × 5 ft 6 in. long, with a ring at each side for the 'cord.'   The 'crook' at the head is shaped like a sickle.   The whole seems to have been fixed on a pole.

[2] *Archæologia Aeliana*, XXII, 1900.

[3] *South Yorkshire (the Deanery of Doncaster)*, I, p. 370.

which means to stake or to pale and 'a stake' and 'a stowre' are both translated 'palus' in the same vocabulary: to its compiler, only stakes were necessary 'to make a howse.'

Such houses were easily made and as easily broken into. In the *Select Pleas of the Crown* of the thirteenth century, reigns of John and Henry III[1], the thieves, whose misdeeds are recorded therein, break either the walls or the doors of the houses indiscriminately. In the *Catholicon* an 'howse breker' is 'apercularius,' and in the same vocabulary 'perpalare,' that is, literally, to get through pales, is one of the translations of 'to thirle,' that is, to make a hole. In the *Gesta Romanorum* there is 'if any thirle or make an hole in a feble wall of a feble house.' Our term 'house breaking' is a survival from the days when thieves did actually break the houses which they entered.

Descriptions have been given of wattlework in Kent and Derbyshire of the most usual type, in which the twigs are woven in-and-out of upright rods. In East Anglia a good deal of wattle and daub seems to have had uprights only, which 'usually consisted of hazel sticks, about an inch thick, with the bark left on,' as in an example from Cross St., Sudbury, Suffolk[2].

In a dilapidated Tudor house at Harlton, near Cambridge, the writer found an interior plastered partition, with studs 4 in. × 6 in. fixed from 2 ft. to 2 ft. 6 in. apart. Each space between the studs was filled with upright straight branches, about ten in number, split into halves and varying in thickness from $1\frac{1}{4}$ in. to 3 in. These were plastered on each side. Such a partition explains the terms 'perpalare' and 'feble house' above. Later work of the same kind about Cambridge has usually *horizontal* split branches, laid over the vertical, but even so, outer walls so constructed are very flimsy.

Somewhat similar was the method used at Saffron Walden, in Essex, in the fifteenth century. 'Long, split saplings of ash and hazel, arranged vertically between each pair of timbers, were sprung into grooves in the upper and lower horizontal beams. They were then tied with green withy bands to short cross-pieces wedged between the uprights. Clay mixed with short straw and chalk was then worked in between the main timbers, thus covering

---

[1] Edited for the Selden Society by F. W. Maitland.

[2] Illustrated by Mr Basil Oliver in *Old Houses and Village Buildings in East Anglia: Norfolk, Suffolk and Essex*, pp. 10 and 37.

the hurdle-like lathing with a rough "daub" cement. Whilst still wet the clay was scored over to give keying to the finishing coat of fine plaster, which was itself coloured and often painted with decorative designs on the inner surface[1].'

More primitive was a partition in a mediæval house, said to have been formerly the parsonage, at St Florence, near Tenby, in English-speaking South Pembrokeshire, which was described to the writer as having 'just brambles' for the foundation for the daub. The uprights of the wattling followed the same development as the rafters, from round to rectangular, and the writer has seen specimens of the latter form in a cottage by St Issell's church, near Tenby, and in a cottage at Cymryd, near Conway, the uprights were about $1\frac{1}{2}$ in. square, and the wattles were interwoven round these vertical pieces[2]. In another wattlework partition in South Wales, which the writer found in a cottage at Marros, Carmarthenshire, the uprights were square but they were fixed horizontally, like plastering laths, with the upright wattling woven around them. This was no doubt due to the influence of contemporary plaster on laths, and in the well-known but abnormal example of wattlework exposed in the Elizabethan house, Plas Mawr, at Conway, the wattle twigs are actually woven in and out of horizontal riven oak laths. Here the laths are about $2\frac{1}{2}$ in. wide and $3\frac{1}{2}$ in. apart : the wattle twigs average $\frac{1}{2}$ in. in thickness, but some are much smaller, and they are woven alternately round the laths in the usual manner, that is one in and the next out, as in basket work. Similar wattlework is to be seen at Penrhyn Old Hall, between Llandudno and Colwyn Bay, which is possibly earlier in date.

In a house at Warrington 'the wattles are rods of hazel, with the bark on, laid close together in an oblique direction and covered with a thick coating of clay and cow dung[3].' Probably the rods were placed in an oblique direction in order to strengthen the wall against the 'howse breker.'

We have now no such buildings as the 'frail houses' at Morpeth, which must have been mere hurdles, like the Romano-British examples discovered by General Pitt-Rivers. Existing wattlework

---

[1] Guy Maynard, 'Some Account of Saffron Walden Museum' in the *Antiquary*, Oct. 1915 (p. 371).
[2] H. Hughes and H. L. North, *Old Cottages of Snowdonia*, p. 19.
[3] S. O. Addy, *Evolution of the English House*, p. 109.

is sometimes in stone, but more usually in timber framed buildings, and the earliest framing was tied together, as it was in Germany also, according to R. Henning[1], who says that the primitive mode of joining the framed building was to bind it with cord like the scaffolding of buildings to-day. And it was so with the English wattlework, for tying was the most primitive mode of fixing it in a framed building. Sir Rider Haggard has described some cottages in Norfolk : he says that these 'at their last reconstruction which I should judge to have taken place one hundred and fifty or two hundred years ago, were largely built of studwork, framed on sapling boughs, measuring about $1\frac{1}{2}$ in. in diameter, and lashed to the roof beams with string,' of which he secured some pieces, and found that it 'was for the most part still fairly sound and of a very strong and even make[2].'

More secure, and therefore more permanent, methods of fixing the vertical round stakes were obtained by fitting their ends into grooves or into round holes in the heads and sills of the wooden framework of the wall. The former is the less advanced method, and for it roughly made grooves were run on the underside of the head and the upper side of the sill: the upright posts or stakes were cut to the required lengths, and then slid along in the grooves and knocked up tight. The construction of this kind of wattlework in South-West Surrey has been described: 'The frame being put together of oak, the panel is formed by fixing hazel rods upright in grooves, cut in top and bottom, and by then twisting thinner hazel wands round them. The panel is then filled solid with a plaster of marly clay and chopped straw, and finished with a thin coat of lime plaster. The same system is used for inside partitions and occasionally the lattice is formed of oak laths[3].'

In South Yorkshire there was an intermediate form in which the upper ends of the stakes were fixed in holes or posts and the lower ends in grooves. Occasionally the studs of lath and plaster partitions in South Yorkshire were fixed in this manner, and at the house at Cymryd, near Conway, already mentioned, the uprights were wedged into holes along the sides of the principal and the underside of the upper tie, by being driven along grooves on the

[1] *Deutsche Haus*, p. 164

[2] *A Farmer's Year*, p. 325.

[3] Ralph Nevill, *Old Cottage and Domestic Architecture in South-West Surrey*, p. 19.

upper sides of the two ties[1]. A better method was that in which each end of the stakes was fixed in a hole. In wattlework of this kind in a house, now ruined, at Cranemoor, South Yorkshire, the writer found that the stakes were fixed at an average of 18 in. centres, the stakes were 2 in. in diameter, the holes were $1\frac{1}{4}$ in. in diameter and 3 in. deep, and the ends of the stakes were whittled down to fit into the holes. In a cottage at Somersall-Herbert, Derbyshire, one of the cross beams 'is pierced on its under side with round holes about $\frac{1}{2}$ in. or $\frac{3}{4}$ in. diameter, and about 15 in. to 18 in. apart,' evidently for former wattle stakes. All these varieties of wattlework were designed to be put up without the use of iron nails, which were too expensive to be used, except with parsimony, in the old ordinary buildings.

The general and widespread use of wattle and daub in England is shown by its varied names in the dialects. In the West of England wattlework was known as 'freeth' or 'vreath,' and the writer has heard it so called in the English-speaking part of Pembrokeshire, in South-West Wales. The word is perhaps a variant of 'wreath.' In the year 1436 the churchwardens of St Michael's, Bath, paid two pence for 'stodyng et frethyng,' and in the year 1441 they paid six pence for twigs, and then two shillings for 'vrethyng et dawbyng.' Later they paid for 'staking et frethyng,' and later again, in 1535–6, they bought 'frithyng roddes[2].' The churchwardens of Yatton, Somerset, paid two pence in 1446 for 'ryses for the daubes.' The sticks used for wattlework in East Anglia are known as 'rizzes' or 'razors,' and 'rice' is a widespread dialect word for brushwood and small sticks. In Scotland, wattle and daub is called 'stake and rice,' the rice being brushwood. However, it is possible that the 'ryses' at Yatton were rushes, for rush and plaster partitions were common, and a South Yorkshire example, recently destroyed, was described to the writer by its destroyer as made of wide laths, three or four inches wide, fixed twelve or sixteen inches apart, with rushes woven in and out, and the whole covered with clay. The writer has been informed by an old Carnarvonshire carpenter that in the Lleyn district, interior partitions used to be made of rushes or straw woven into ropes and twisted in and out of uprights, after the manner of wattlework.

Later mediæval accounts relating to the 'house on the bridge'

---

[1] H. Hughes and H. L. North, *Old Cottages of Snowdonia*, p. 19.

[2] *Proceedings of the Somersetshire Archæological and Natural History Society*, 1877–80.

at Bridgwater, in Somerset, show that it was daubed, and had 'spekis and yerdis,' that is, wattle twigs and rods[1]. The West Country 'yards' are the same as those in King Alfred's *Blossom Gatherings from St Augustine* before mentioned.

The old oaks of Staverton Park have been subjected to frequent 'stowings' to provide 'stallons' or 'wattle sticks' for the erection of wattle walls plastered with mud[2]. In Essex, the terms 'splint' and 'stovett' are used for the 'pieces of wood used in making wood houses plastered with clay,' and they are so used in the churchwardens' accounts of Great Bromley, in the year 1627.

Wattle and daub is called 'stud and mud' in Lincolnshire, and in the northern part of the county 'daub and stower' also. It is called 'rice and stower' in the North country, the rice being brushwood as we have seen. In Cheshire and Lancashire it is called 'raddle and daub,' and this is shortened into 'rad and dab.' In the Peak of Derbyshire, according to Mr Addy, mortar is said to be too rad when it is too porous and loose, and contains too little lime and too much sand[3], and in Kent a 'raddis chimney' was a chimney of wattle and daub. The term is also shortened into 'rab and dab,' and used for a form somewhat like the stud and mud of Denmark, in which the mud is filled in between the woodwork. 'Raddle' is probably cognate with 'hurdle,' as *Withal's Dictionary* of the year 1602 gave 'paries craticus' as a translation of 'a hartheled or ratheled wall.' And both these words are perhaps cognate with the widespread dialect word 'wrastle,' the weaving of the wattlework in-and-out of the stakes[4].

It has been shown that there is a form of wattle and daub wall in which the wattle is of two thicknesses, with the daub and mud packed between them. According to Professor Sullivan, in ancient Ireland 'round houses were made by making two basket-like cylinders, one within the other and separated by an annular space of about a foot, by inserting upright posts in the ground and weaving hazel wattles between, the annular space being filled with clay. Upon this cylinder was placed a conical cap, thatched with reeds or straw. The creel houses of many Highland gentlemen

---

[1] *Proceedings of the Somersetshire Archæological and Natural History Society*, 1877–80.

[2] V. B. Redstone, 'Norman Rule and Norman Castles,' in *Memorials of Old Suffolk*, p. 40.

[3] *Notes and Queries*, 10th Series, II, p. 202.

[4] See also note on p. 127.

in the last century were made in this way, except that they were not round[1].' The making of the old stud and mud cottages at Great Cotes, in Lincolnshire, was described by Mr John Cordeaux : they were formed by driving in rows of stakes and then trampling in with the feet, between the stakes, wrought clay mixed with chopped straw. The crofts and paddocks of the village were fenced with the same kind of walls in the middle of the eighteenth century[2]. This form of construction also shows itself in North Wales in round banks of earth which indicate former round houses, as may be seen on the slopes of Moel-y-Gest, between Criccieth and Port Madoc.

We have now seen how widespread was the use of wattle in the buildings of England, and the wattlework walls may be grouped into four principal varieties as follows :

(1) wattlework only,

(2) single wattlework daubed on one side only,

(3) single wattlework daubed on both sides,

(4) double wattlework, filled in between with mud or other materials, as in the bark peelers' hut of High Furness, and the examples mentioned above.

Of these varieties, neither (1) nor (2) is now found in England as far as the writer is aware, and (3) is the most usual form.

The solidity of the filling of the fourth and last type would show to the builders, probably by the natural decay of the wattlework exterior, that the latter was not a necessity, and thus would be developed the buildings of mud which are widespread over England, Wales and Ireland under various names, such as 'clom' and 'cob,' which are dialect names for mud walls, of which the use in ordinary buildings was once widely spread in England. The English student of architectural history when reading of the mud architecture of Babylonia, thousands of years ago, is apt to overlook the comparatively recent mud buildings of his own country. 'Cob' is a West of England word, and its literary history is short. The older form seems to be 'clob,' for a seventeenth century Devon inquest speaks of 'a mudde or clobbe wall[3].' Whether it is derived from 'clob' or from 'korb,' the evidence in either case is in favour of a wattlework origin for the mud wall,

---

[1] *Encyclopædia Britannica*, 9th edition, XIII, p. 256.
[2] *Notes and Queries*, 5th Series, III, p. 487.
[3] Ibid. 11th Series, I, p. 426.

as 'korb' means basketwork, and 'clob' and 'cleam' are forms of widespread Teutonic words for smearing on and plastering. In the complaints which are made from time to time of too onerous building bye-laws in rural districts, cob is sometimes cited as a cheap and desirable form of walling, of which the use is forbidden. It is probable that no walls depend for their durability so much on the nature of the material as do those of mud. If the mud or cob be of the right kind, there seems no good reason why it should not be permitted for the walls of houses in rural districts, for a wall made of cob which contains proper lime may be regarded as built entirely of mortar. A Sheffield joiner has told the writer that in the middle of the last century, about Banbury in Oxfordshire, his early home, the men who worked on the roads built their own cottages of road scrapings, layer by layer, and when these were set it was almost impossible to pick the walls down. According to Mr C. E. Clayton[1], the walls of the mud-built cottages in the New Forest were formed of clay mixed with chopped straw or stones, packed down between boards or hurdles, these latter, unlike those of the bark peelers' huts, being removed when the clay dried and hardened. Monsieur C. Enlart says that this kind of walling was much used in some southern portions of France, and has left examples since the sixteenth century. On the contrary, in the North of France, in all the region which now makes use of brick, daub was formerly much used. No ancient example has come down to us, but this kind of building still remains frequent in Ponthieu and in a part of Artois[2].

Mud walls were in use for London houses as early as the year 1212, and in the London Assize of that year there occurred the word 'torchiator' which is cognate with the French word 'torchis,' that is, mud, from which is derived the modern word 'torched[3].' In the year 1809 in Leicestershire, road scrapings were considered to make the best mud for the walls of cottages and such cottages were often built and thatched by the labourers themselves. The mixing of the mud with straw and stubble was called tempering[4].

---

[1] 'Cottage Architecture' in *Memorials of Old Sussex*, p. 291. (See also Fig. 51.)

[2] *Manuel d'Archéologie Française*, II, pp. 185 et seq.

[3] T. Hudson Turner, *Domestic Architecture in England from the Conquest to the end of the Thirteenth Century*, p. 281.

[4] William Pitt, *General View of the Agriculture of the County of Leicester*, p. 26. The *General View* for Dorset also says that road scrapings were used in that county for cottage walls (p. 85).

Mr S. O. Addy has described a mud house at Great Hatfield, Mappleton, in the East Riding of Yorkshire. He says, 'The walls are built of layers of mud and straw which vary from five to seven inches in thickness, no vertical joints being visible. On the top of each layer is a thin covering of straw, with the ends of the straw pointing outwards, as in a corn stack. The way in which mud walls were built is remembered in the neighbourhood. A quantity of mud was mixed with straw, and the foundation laid with this mixture. Straw was then laid across the top, whilst the mud was wet, and the whole was left to dry and harden in the sun. As soon as the first layer was dry another layer was put on, so that the process was rather a slow one. Finally the roof was thatched, and the projecting ends of straws trimmed off the walls. Such mud walls are very hard and durable, and their composition resembles that of sun-burnt bricks[1].'

This method of making walls of mud, described by Mr Addy, is typical of that generally used in England, as for instance in Buckinghamshire, where the walls were built of a kind of white clay called 'witchit,' found about eighteen inches below the surface of the ground. But walls of mud had their peculiar home in the West, and the volumes which treat of the Western counties in the survey of the state of agriculture in the counties of England published by the Board of Agriculture and Internal Improvement about a hundred years ago, contain descriptions of its contemporary use in building. Wet was a great enemy of cob walls, and in order to resist it and also the inroads of vermin, it was necessary to lay a foundation of stone. The cob should project over the stone foundation about 1½ in. There was a Devonshire saying that all cob wants is a good hat and a good pair of shoes, that is, a stone foundation and a coping of thatch: the latter, a usual protection for cob in the eighteenth century, was used in the Middle Ages as a temporary protection for the unfinished ashlar walls of great churches and similar buildings.

Cob is very durable when properly protected, and in the middle of the last century in Devon, there were houses in perfect preservation which had been built in the reign of Elizabeth.

In building the walls the only implements used were a dungfork and a 'cob parer,' which was like a baker's peel, the shovel for

---

[1] *Evolution of the English House*, p. 40. Here, and elsewhere, the mud walls are damaged by bees, which sink their nesting holes in them.

removing bread from the oven. The first layer of cob was built
2½ ft. high all round the foundation and the walls themselves were
2 ft. thick. The stuff was used as wet and soft as ordinary mortar,
and after a week or so, according to the dampness or dryness of the
atmosphere, allowed for the layer to consolidate, another layer was
put on, and so on until the work was finished, two years being
required for a two storied house if it were to be properly done. In
Devonshire each course was known as a 'raise.' When building a
cob wall one of the workmen stood on the wall to tread it down,
and the woodwork, such as the lintels of doors and cupboards, was
fixed in as the work went along. The walls had a tendency to crack,
especially at the corners, and they were generally rounded to avoid
this: but this rounding of the corners may have had its origin in
early circular or oblong buildings, and been retained for practical
purposes. The cob also scaled off and bulged when the whitewash
or plaster with which it was usually coated became decayed, and
thus some Devon villages had an extremely dilapidated appearance
a century ago. Charles Vancouver, the writer of the Devon volume
in the *General View*, published in the year 1813, complained that
it was impossible to distinguish a village of cob from a bean field,
and he noticed with pleasure many rough cast and whitewashed cob
cottages in the neighbourhood of Exeter[1]. It was an age impelled
to utilitarianism by grim and insistent necessity, and the fitness of
the building for the landscape did not appeal to the writer.

Where there was chalk, as in parts of Dorset, it was ground up
with the cob to its improvement. Chalk was easily obtainable in
Hampshire, and in the year 1810, according to Charles Vancouver[2],
the cob there was composed of three parts of chalk and one of
clay, well kneaded and mixed together with straw. Where the
chalk was not easily obtainable it was only used for the rough cast
or the finishing coat, after being ground in a circular trough. In
sandy and heathy districts of Dorset, loam, gravel and sand were
used, and heath was used as a binding material instead of straw,
but such cob was not as durable as that composed of chalk[3]. In
Cornwall cob was composed of two loads of clay to one load of
shilf, that is broken slate in small pieces such as is used for
mending roads, barley straw was added afterwards. Each 'raise'
was diminished in height.

[1] p. 26.   [2] *General View of the Agriculture of Hampshire*, p. 67.
[3] *General View of the Agriculture of the County of Dorset*, p. 86.

In cob buildings in West Somerset the gables were made of rough round poles or sticks nailed upright, and across these some split sticks were nailed to serve as laths, and over all was a coat of daub, or very rough mortar: such work was known as 'split and dab[1].'

These split sticks of Somerset show how narrow is the boundary between laths and wattles, and the transition from wattle to lath work was probably in the weaving of the split laths in and out of the uprights or studs, after the manner of some present day basket work. Mr C. E. Clayton has described and illustrated such interlaced laths from a 'very early cottage at Steyning[2].' Similar work has been described by Mr F. W. Troup: he says, 'I have recently had to make some repairs and alterations to a beautiful example of a half timber house in Norfolk. It was entirely framed of oak on a low brick plinth about 15 in. high and, curiously enough, the timbers showed inside the house, but outside not at all....The filling in between the oak timbers was done by means of oak poles about as thick as one's wrist, wedged in upright between the horizontal timbers and secured there, and thin laths were woven in to give a key for the clay daub, which formed the solid part of the filling in. This wattle and daub varied in different districts, but it was the usual method of filling the framework, and the clay was scratched on the surface for a key, and then sometimes coated with the merest skin of lime plaster, as better able than clay to resist rain and wind. Clay, however, had its own use, and I doubt if anything better can be devised as a filling-in material for timber framed houses[3].'

In work of superior quality riven laths have been used in England from early times. Professor Thorold Rogers found, from the building accounts, that mediæval laths were generally of oak or beech, and that they were distinguished in the accounts as 'sap and hart' and 'hart and sort.' They were also classified according to their strengths, as at the present day. Sap laths as a trade term continued in use to the end of the eighteenth century. The accounts of the rebuilding of Bolsterstone Chapel in South Yorkshire in 1791, printed by Mr J. Kenworthy, in his *Early History of*

---

[1] *English Dialect Dictionary*, *s.v.* Split. See also p. 58 ante.

[2] 'Cottage Architecture' in *Memorials of Old Sussex*, p. 295.

[3] 'Influence of Materials on Design in Woodwork' in *Arts Connected with Building*, p. 68.

Stocksbridge and District include thousands of 'sap lats[1].' Rogers considered that the rending of laths in the Middle Ages was a by-industry, and that it was carried on in the woods[2].

By the statute of Edward III, a heart of oak lath was to be one inch wide and one half inch thick, but in practice it became reduced to a bare quarter of an inch. The Assize of Bread in the year 1528 declares that 'the lath shall conteyne in length v fote, and in brede ii ynches, and in thyckenes halfe a ynche of assyse upon payne for every c lathe put to sale to the contrarye, iid[3].' Some of the old laths were very broad, for example, at a house in the Hope Valley, Derbyshire, which was about three hundred years old, the laths of the plasterwork were about 4 in. broad[4]: such laths were possibly made from the staves of disused casks or barrels, just as were the palings of great houses in the thirteenth century, and in country places as lately as the nineteenth century, according to the *Dictionary of the Architectural Publication Society.* Monsieur C. Enlart states that the laths which are nailed on mediæval timber framing ('pan de bois') in France are often made of the staves of ancient barrels[5].

Early sixteenth century laths in the Sheffield district were fixed by 'shooting,' that is, they were sprung into V-shaped grooves cut into the sides of the upright studs. Occasionally, as at the now ruined Oxspring Lodge, near Penistone, the studs were inclined (Fig. 47). The laths were fixed a little distance apart to give the usual key, and as these were 'full long' their spring kept them in position. This method of fixing the laths recalls some of the early wooden partitions, and one of the methods of fixing wattle uprights already described, with the difference that the grooves for the laths are vertical, while in the previous examples they are horizontal. The Swedish 'skiftes værk,' in which the laths are replaced by solid deals without plaster, is more nearly allied.

At the same period, in South Yorkshire, an attempt was made to provide a stronger panel filling than the usual oak board, by the use of the local stone or 'grey' slates, which were fixed between the

---

[1] xv ('Bolsterstone'), p. 56.

[2] *History of Agriculture and Prices*, I, pp. 487–8, IV, pp. 435–6. This history, not very happily named, contains much information as to prices of materials and wages in the building trades during many centuries.

[3] *Dictionary of the Architectural Publication Society, s.v.* Lath.

[4] Information supplied to the writer by Mr J. R. Wigfull, A.R.I.B.A.

[5] *Manuel d'Archéologie Française*, II, pp. 185 et seq.

studs by being fitted into the grooves in their sides, and then plastered over. Such stone slate filling was used on the ground floor
at Oxspring Lodge, and the 'shot' laths on the upper floors
(Figs. 47 and 48).

In a partially ruined house near Firth Park, Sheffield, the

Fig. 47.   Oxspring Lodge, near Penistone.   The studs are inclined
on the upper storey.   The original laths were not nailed on the
studs but 'sprung' in grooves on the edges.   The lower studs
were filled with thin stones shown in Fig. 48.

writer found that the lath and plaster wall had been filled with
straw fragments, and the oldest stone built cottage in Snowdonia
known to Messrs Hughes and North had the centre of the walls,
which are usually left dry, filled with bran[1]. The non-homogeneous
wall is of great antiquity.

[1] *Old Cottages of Snowdonia*, p. 13.

The materials used for daub and plaster were closely similar to those used for mortar, and the statement previously made as to lime in mortar also applies to its use in plastering. It is said that earth of any kind that would set was used for the 'stud and mud' of Leicestershire, and at Skirlaugh, in that county, all the cottages built before the nineteenth century have stud and mud walls. The walls of the outshot of a cottage at North Meols, West Lancashire,

Fig. 48.   Partition on the ground floor of Oxspring
Lodge, Penistone. Thin roofing stones, known
locally as grey slates, are fixed in grooves on
the edges of the upright studs.

have been described by Mr Addy. 'The posts or studs stand close together, and the walls are plastered by clay mixed with straw to the depth of an inch, the plaster being covered by several coats of white lime. This kind of building is known in Lancashire as "clam, staff, and daub[1]."' The clay used for these clay walls was trodden by men with their feet, and mixed with the star grass which grows on the sand hills[2]. In South Yorkshire

[1] *Evolution of the English House*, p. 44. There is an illustration of the outshot on the preceding page (43).
[2] Ibid. p. 47.

the writer finds that much old plaster is mixed with straw, with lime in the finishing coat only. Before the improvements in the means of communication which took place during the nineteenth century, the carriage of lime into districts which did not produce lime was difficult, and a very old Derbyshire man once described to the writer the lines of horses which he had seen crossing the moors, in single file, and carrying panniers of lime from the Peak of Derbyshire to Sheffield. In South Yorkshire the straw was trodden into the plaster by women, and in East Lancashire by boys. In East Anglia and in Cheshire horses were used for the purpose: in the latter county the clay and straw were placed in a farmyard, and the horses trod it until it was thoroughly softened and moistened. In the year 1477–8 the churchwardens of St Michael's, Bath, paid for hay and straw for daubing ('ffeno et stramine[1]'). The earliest use of the word daub, as given in the *Oxford English Dictionary* is from the year 1325, viz., 'Cleme hit with clay comely within and alle the entendur dryuen daube withouten.' The parish of Ludlow possessed a church house which was often in need of repair, for which purpose clay was used and the accounts show that in the year 1537 the churchwardens bought 'roddes to wynde ii walles in the churche howse.' Palsgrave wrote in the year 1530 that 'daubing may be with clay onely, with lime plaster, or lome that is tempered with heare or strawe.' Harrison in his description of England in 1586 wrote, 'The claie wherewith our houses are empanelled is either white, red or blue,' of which the first partook of the character of chalk and the second of loam[2]. According to the *Builder's Dictionary*, 1734, 'Loam is a sort of reddish earth, used in buildings (when tempered with mud, jelly, straw, and water) for plaistering walls in ordinary houses[3].' In the *Old Halls of Lancashire and Cheshire* the plaster panels of the buildings are said to be 'filled in with a basket work osier foundation, daubed over with clay strengthened with stringy weeds. The finishing coat is of plaster on both sides, richly matted with hair, and frequently set back half an inch or more.' The materials for daubed walls were beaten up and incorporated with a sort of club called a 'dauber's beater.'

---

[1] *Proceedings of the Somersetshire Archæological and Natural History Society*, 1877–80.
[2] In the chapter ' Of the Manner of Building and Furniture of our Houses,' which is an interesting contemporary description of Elizabethan buildings, and should be read.
[3] *Dictionary of the Architectural Publication Society*, *s.v.* Loam.

When stone walls superseded wattle in North Wales the stones were at first daubed like the wattle with a mixture of clay and cow dung[1]. This had a wider range than North Wales and in England such important buildings as the churches were often daubed or plastered in the Middle Ages. As an instance the church of St Michael, Bath, in the year 1394, was daubed with lime and sand both inside and out ('tam infra quam extra[2]').

According to Mr Bankart the English-made plaster which succeeded the stucco-duro of the Italians in East Anglia, contained a certain amount of cow dung and road scrapings[3]. Batty Langley,

Fig. 49. Cottages at Scarrington, Nottinghamshire. The walls are about two feet thick constructed of slabs of dried mud, plastered over with mud.

in his *London Prices*, 1750, states that in pargetting about one-fourth part of dung was used which was incorporated with the lime by well beating it[4]. In this the English plasterers were only following the English tradition, for the use of cow dung is old and widespread, and it has excellent setting properties. It is still used in out-of-the-way parts of North Derbyshire as a material for parging flues and chimneys. R. Holme in his *Academy of Armory*, published in 1688, calls mortar mixed with dung for flues 'pergery

---

[1] H. L. North, *Old Churches of Arllechwedd.*
[2] *Proceedings of the Somersetshire Archæological and Natural History Society*, 1877–80.
[3] *Art of the Plasterer*, p. 57.
[4] *Dictionary of the Architectural Publication Society*, *s.v.* Parget.

mortar.' He also writes of mortar mixed with small stones to fill the middle of a wall[1]. A German saying is quoted in J. and W. Grimms' *Deutsches Wörterbuch* that where one has no chalk one must daub with dung ('mit Drecke kleiben[2]'). In some cases the daubing of dung was periodically removed for fuel, thus, Naseby, in Northamptonshire, in the year 1792[3], with the exception of a few of the modern and best houses, was built principally with a kind of kealy earth, very durable, and capable of lasting two hundred years (Fig. 50); but, the vicar complained, instead of being drawn over with lime mortar and marked or limed to appear

Fig. 50.   Old cottages at Naseby, Northamptonshire.   The walls are of mud, and the roofs are covered with straw thatch.

as stonework, which might have been done at a moderate expense, the new coat, which they had once a year, consisted of cow dung spread upon them to dry for firing.   Probably this was an ancient custom as the use of dung for fuel is very widespread.   In the memory of a writer living in 1893 the house-walls in the Isle of Portland 'looked passing strange with dung stuck on them to dry[4],' and the mud walls of houses in the United Provinces of Upper India are usually decorated with patches of cow dung stuck on to dry in the sun, and then used for fuel[5].

[1] p. 262.   [2] *s.v.* Kleiben.
[3] Rev. J. Mastin, *History and Antiquities of Naseby*, p. 7.
[4] H. J. Moule in *Notes and Queries*, 8th Series, IV, p. 377.
[5] Lt.-Col. S. J. Thomson, *Silent India*, p. 27, and *Real Indian People*, p. 70.

Naseby is said to have consisted almost entirely of mud buildings until about 1850. The walls were built in layers in the usual manner, and each layer was beaten with a sort of fork by a man standing on the wall, and then left to dry. Mud is not used now except occasionally for garden and field walls[1]. Other old mud buildings are shown in Figs. 49 and 51.

In the plaster panels of timber framed buildings the face of the plaster was, in the oldest examples, set back from the timber

Fig. 51. Old cottages on St Catharine's Hill, Bournemouth. The walls are of mud, plastered over. At the left side of the photograph the plaster has fallen away, showing the mud wall.

framing; then, in later work, the face of the plaster was made flush with the timber work; and, finally, the timber work itself was plastered over, the plaster being 'flush' over panel and framing alike. Mr F. W. Troup, writing of an old Norfolk house as already mentioned, says the house 'For some reason—and it appeared to

[1] J. B. Twycross in *Country Life*, 14th March 1914. Other descriptions with illustrations of mud buildings in England appeared about the same time in *Country Life*. Figs. 49, 50 and 51 are reproduced by the permission of the proprietors.

be original too—had had the oak timbers all hacked over outside to form a key for plaster, and the whole surface was plastered over, oak and panels alike, forming a plastered house[1].' Such work was laid on with a float or other tool, but daub was thrown at the wattle, and this seems to have been the recognised method from East Anglia to South Pembrokeshire. Rough-cast is now thrown and not laid on, and like the old method of daubing is probably a survival from the time when the plasterer's tools were his hands. In the year 1519, Wm Horman wrote in his *Vulgaria*, 'Some men wyll have theyr wallys plastered, some pergetted, and whytlymed, some roughe caste, some pricked, some wrought with playster of Paris[2].'

The workman who constructed the walls was known as a wall-wright in the tenth and eleventh centuries, according to the contemporary glossaries, and the word is a usual translation of 'cementarius,' that is, a worker in plaster or daub. In another eleventh century gloss 'cementarius' is translated by stonewright as is 'lathomus' by Aelfric in the preceding century[3]. The word 'dealbabor' is translated in a tenth century gloss as 'ic beo gehwitad,' that is, I am whitened. The word 'dauber,' widely spread in England, is derived through the French from the Latin 'dealbare' meaning to whiten, and according to the *Oxford English Dictionary* does not appear in English before the fourteenth century ('in that cofer that watz clay daubed'). The word was not restricted to whitening walls but applied indiscriminately to the daub or plaster of solid walls, of wattle, or of laths. In the year 1538, Elyot defines 'cementarii' as 'daubers, pargetters, rowghe masons whiche do make onely walls,' and the modern term 'stonemason' implies that there were once masons who made walls of other materials than stone. In Devonshire the workmen who constructed buildings of cob were known as cob masons. According to the occasionally vague *Dictionary of the Architectural Publication Society* 'in the West of England, the mason that sets the stone is called a "rough mason."' Moxon, the contemporary of Wren, in his *Mechanick Exercises*, distinguished between the white mason who hewed stone, and the red mason

---

[1] 'Influence of Materials on Design in Woodwork' in *The Arts Connected with Building*, p. 70.

[2] p. 241.

[3] Wright-Wülcker, *Vocabularies*, s.v. Cementarius and its variants.

who hewed brick. In the year 1544 the churchwardens of St Giles', Reading, paid four pence to a mason for lathing and daubing, and in this case the mason was what we should now call a plasterer.

A faint reflection of the Italian Renaissance is found in some English ordinary plasterwork, such as the external work in relief which is most usual in East Anglia and to which the name pargetting is specially applied[1]. Old rough cast was sometimes sprinkled with bits of old glass, and in Kent it was given a blood red colour by the use of red sand, or a kind of bluish colour by the use of powdered cinders[2].

Plaster is now made of lime, sand and hair, and the use of earth and of animal and vegetable matter is obsolete. Recent developments are in the usual direction of specialisation in material, in the use of patent mixtures prepared by manufacturers. The oldest of the cements was Parker's, Sheppey or Roman, so called because it was believed to be like the Roman, came from Sheppey, and was patented by James Parker in the year 1796[3]. The cement which has superseded all others of its class, and indeed caused a new era in certain spheres of building construction, is that known as Portland, which was invented about ninety years ago. (See also p. 278.)

Other modern developments in plastering are the application of the plaster dry in sheets or slabs, in the form known as fibrous plaster and the substitution of metal lathing for wooden laths, both of which were invented about the middle of the last century.

In a former chapter we saw that a certain village carpenter in the twelfth century was under an obligation to do plumber's work at the church as required[4], and this was probably due to the absence of a plumber in the district. Similar cases in other trades must have been frequent in early times, when expensive materials such as lead and ashlar stone were only rarely used, and competent craftsmen could not be summoned by telephone. One of the Manx stone crosses provides another example, for on it a smith inscribed the boast that he had made it 'and all in Man.' The boast was untrue, as there had been crosses in the island for hundreds of

---

[1] See Mr Bankart's *Art of the Plasterer* and Mr Oliver's *Old Houses and Village Buildings in East Anglia*.

[2] *Dictionary of the Architectural Publication Society, s.v.* Rough cast.

[3] Ibid. *s.v.* Parker's cement.   [4] Ante pp. 94 and 95.

years : its interest lies in the evidence that a smith did the work of a carver in stone.

Before the Norman Conquest stone was apparently dressed with a double axe, for 'bipennis' and 'bipennis securis' are translated by 'twibille,' 'stanaex' and 'twilafte aex' in contemporary glossaries[1], and the early glossaries also show that the walls were 'tried up' to keep them perpendicular. In an Anglo-Saxon glossary of the eighth century 'perpendiculum' is translated as 'colthred pundur,' and in Aelfric's glossary of the tenth century it is translated by 'wealles rihtungthred,' and in a glossary of the eleventh century it is 'walthraed thaet is rihtnesse[2].' Later, in the fifteenth century *Catholicon Anglicum*, a 'levelle' is translated by 'perpendiculum,' and this is glossed as a 'plommett,' and in the same dictionary a 'swyrre' (that is, a carpenter's square) is translated 'amussis' and 'perpendiculum,' and by Baret 'amussis' and 'perpendiculum' are translated 'leauell line or carpenter's rule.' Huloet translates 'level or lyne called a plomblyne, perpendiculum.' In the *Promptorium Parvulorum* a 'plumbe of wryghtys or masonys' is also translated 'perpendiculum.'

Wattles, and daub, and mud are used no longer for the walls of ordinary buildings, and their place has been taken by brickwork. Although it is now in general use in England, the early history of brickwork is more obscure than that of any other modern form of wall or partition. Lacing courses of bricks were used by the Romans in Britain in their stone walls ; such Roman bricks were approximately square and only about 1 in. thick, and thus they were what we should now call tiles rather than bricks. Similar tiles or bricks were used by Saxons and Normans in their buildings, principally at or near Roman sites, and it is generally considered that they are Roman materials reused. St Martin's church at Canterbury, and St Alban's cathedral are well-known examples, the one Saxon, the other Norman. In the neighbourhood of the Roman wall, in Northumberland, the Roman buildings do not contain lacing courses of brick, nor do the pre-Conquest buildings built of the Roman materials in the Roman manner, such as the chancel of Jarrow church (Fig. 70). At St Patrick's chapel, Heysham, Lancashire, thin stones are used in a rough imitation of the courses of bricks of Roman walls, and this is some evidence that

[1] Wright-Wülcker, *Vocabularies*, s.v. Bipennis.
[2] Ibid. s.v. Perpendiculum.

the pre-Conquest builders of the chapel could not make burnt
bricks.   The glossaries show that the Saxons understood the word
'tiles' to mean thin stones.   In Aelfric's vocabulary of the tenth
century, 'tegulae,' the Mediæval Latin word for roof tiles, is
placed with varieties of stones and translated 'hroftigla' (roof-
tiles): probably he was thinking of such fissile stones as those of
Colley Weston, and Aelfric translates the Mediæval Latin word
'tessellae' (floor tiles) as 'lytle fetherscite florstanas,' that is, little
four-cornered floor stones[1].   However in Modern Danish a brick
is known as a tile-stone ('teglsten').

The baking of bricks does not seem to have been practised in
England in the earlier Middle Ages.   There are a few buildings
of burnt bricks of the thirteenth century such as Little Wenham
hall, and St Nicholas' chapel, Coggeshall, in Suffolk, and Caister
castle, in Norfolk, but it is probable that the bricks used in them
were importations.   Professor Thorold Rogers found that bricks
appeared in the mediæval building accounts at the beginning of
the fifteenth century and that they were used in the Eastern
counties long before they were in other parts of the country[2].   At
first burnt bricks were used as a substitute for stone and hence, as
might be expected, the oldest are to be found in East Anglia, a
country peculiarly lacking in good building stone.   In Essex the
mediæval brickwork in mullions, etc., was covered with a fine
plaster, either to harmonize with the older stone, or to imitate
stone; it was the custom to plaster, more or less indiscriminately,
both the inside and the outside of buildings, and this was of
early origin; and it was thought no shame to imitate stone
quoins and window dressings by plastering over them[3].   Buckler
also mentions the custom of covering fifteenth century moulded
brick windows at the time of their construction with a layer
of fine plaster about $\frac{1}{8}$ in. in thickness to resemble stone.
Mr Godman pointed out that the brick mullions in the porch
window of Bures St Mary Church are less than the stone
mullions below, which was probably done to allow of subsequent
plastering.

Professor Rogers found also that bricks were very costly for long

---

[1] Wright-Wülcker, *Vocabularies*, *s.v.* Tegulae and Tessellae.

[2] *History of Agriculture and Prices*, IV, p. 434.

[3] Ernest Godman, *Essay on the Characteristics of Mediæval Architecture in Essex*,
p. 13.

after their reintroduction in the later Middle Ages[1], and as a consequence they were not employed in those ordinary buildings of which the construction is described in this book. Gradually the use of bricks spread. In the year 1437 William Weysey a 'brikemaker' was given powers to make bricks for Speen Abbey[2], and the first payment for bricks in the *Records of the Borough of Leicester* (Bateson) was made in the year 1586–7. Wm Horman in his *Vulgaria* of the year 1519 writes of walls 'of stone or of brycke or of claye with strawe or mudde[3].' The Great Fire of London, in the year 1666, came at the close of a period when men's minds had been filled with new ideas on all subjects, religious, political, and social, when the middle class was increasing in importance and obtaining, as of right, things which had before only appertained to the aristocracy. The Fire seems to have made an immense impression and was probably one of the chief causes of that abandonment of timber construction which took place in the latter half of the seventeenth century. In the rebuilding of London after the Fire, the use of timber was replaced by that of burnt brick. Happily, Sir Christopher Wren was available to turn his genius to the solution of the problem of the satisfactory treatment of the new material and this was especially the case with buildings with stone dressings in the Renaissance style. Moxon, the contemporary publisher, included a treatise on bricklaying in his *Mechanick Exercises*; brick became fashionable, and it was used, as a fashion, in places where stone was both abundant and excellent. In Sheffield there are buildings, put up generations after Wren's time, with walls of rough or rubble stone, faced with brickwork.

Early in the nineteenth century the craft of the bricklayer declined, as did so many others, but it improved again with the so-called Queen Anne revival, commenced by George Devey, Eden Nesfield and Norman Shaw. These architects and their followers studied the older English use of brickwork, especially in the counties around London, and based their work upon the old methods, but an exotic influence was also brought in by such a book as Street's *Brick and Marble of the Middle Ages in Italy*. The results are to be seen to-day in a brick renaissance widely spread throughout England.

[1] *Industrial and Commercial History of England*, p. 11.
[2] *Notes and Queries*, 10th Series, VIII, p. 465, quoting the *Patent Rolls*, 1436–44, p. 145.
[3] p. 240.

Gradually building in brick spread from the Eastern districts where it was first used. In the nineteenth century the use of the material was extended into nearly all parts of England and it is now in general use in the West Riding of Yorkshire, the land of fine sandstones. Similarly the new colliery villages which are being built everywhere on the magnesian limestone of Yorkshire, Nottinghamshire and Derbyshire, ancient in use and honoured in adoption as a building stone, are built of brickwork. The secret of its success lies in its durability, in the ease with which it is handled, and in its cheapness, which arises from standardisation and the facility of its manufacture. For facility of production is the parent of cheapness, and the height of perfection in an age of manufacture by machinery.

The student of building construction early learns that there are two principal 'bonds' used in brickwork, the one named Flemish, the other English. Both of them were used by Inigo Jones, and their use in the seventeenth century seems to have depended on the use of special bricks for facing, or otherwise. It was cheaper to use the Flemish bond, when a special and more expensive brick was used for facing, especially in the thick walls of the seventeenth century, because the proportion of stretchers is greater in Flemish bond. Thus at the Writing Hall of Christ's Hospital, built in 1682, the decorative part of the façade, of red brick, showed Flemish bond, and the other part English bond[1]. It was only natural that Flemish bond, at first used with the better bricks, should become fashionable and in the course of time be used with all kinds of bricks. Flemish bond is rarely to be seen in the country from which it takes its name.

There were two other minor forms of bond, the one with nearly all headers, or perhaps snap headers, and hence known as 'heading bond'; the other, chiefly stretchers, is known as 'flying' or 'Yorkshire' bond, and was used at Little Wenham Hall[2].

The early builders in brick were loath to abandon timber, the traditional English material, and as the *Dictionary of the Architectural Publication Society* says, 'until the year 1835 it was the

---

[1] *Dictionary of the Architectural Publication Society*, s.v. Flemish bond.

[2] Much of the charm in the appearance of old brickwork is due to the irregular bond, almost as much as to the colour of the materials. In some buildings, the bricklayer seems to have been quite content if his bricks merely 'broke joint,' and headers and stretchers are mingled in a delightful manner, quite at variance with the stiff diagrams of modern books of building construction. The thicker walls made this possible.

custom of all good builders, especially where there was any irregularity in the character of the sub-soil, to place one or more lines of balk timber along the centre of the wall at its footings,' and it may be added, smaller ' bond timbers ' as they were called, were placed in the walls also, at various heights. The timber was liable to decay and the wall was then weaker than it would have been without the bond. The elder Brunel, the civil engineer, devised a far more effectual bond by laying hoop iron in the courses of brickwork so as to form a continuous longitudinal bond not liable to decay. ' It was in the Autumn of 1835 that he built, and invited his friends to see, what he called a brick beam, which consisted of seventeen courses of brickwork in cement, having a clear bearing of 22 ft. 9 ins. and having its five lower courses built with twelve lines of iron hooping. This brick beam stood for two years with 25,000 lbs. suspended at its centre, an object of great interest to a number of scientific and practical men[1].' The use of hoop iron became so usual that the Dictionary was moved to say that it formed an important ' epoch in the history of the art of building.' The use has been discontinued for years, except in situations where considerable strength is required in the wall.

The decreased thickness of walls, permitted by the regular size of the bricks led to the invention of hollow or cavity walls, in which the walls are double. Their use is at least one hundred and fifty years old, and they are now very generally used, as they are effective against the intrusion of the weather.

In the past it was more usual to regulate the sizes and qualities of merchandise by law than it is now, and the manufacture and sizes of bricks were so fixed. The bricks at Little Wenham Hall are $9\frac{3}{4}$ in. long, $4\frac{3}{4}$ in. broad, $2\frac{1}{4}$ in. high and by a statute of Edward IV, in the year 1477, the mould for making bricks for building was to be 9 in. long, $4\frac{1}{2}$ in. broad, and the bricks when burnt were $8\frac{1}{2}$ in. long, 4 in. broad, and $2\frac{1}{2}$ in. high. They were further regulated by Acts of Parliament in 1567–8, and in 1625. An Act of 1729–30 ordered that the bricks when burnt within fifteen miles of London should be not less than $8\frac{3}{4}$ in. by $4\frac{1}{2}$ in. and $2\frac{1}{2}$ in. thick, and that the mould shall be 10 in. by 5 in. by 3 in. in the clear. It stated that the ' stock and place bricks may be burnt in one and the same clamp, but apart,' not intermixed. The Great Fire of London gave a great impetus to brick making, and wood

---

[1] *Dictionary of the Architectural Publication Society, s.v.* Hoop Iron.

fuel became very scarce and dear, as the use of the ashes of mineral coal was then unknown[1].

Moxon in his *Mechanick Exercises* said that 'common English bricks' were 9 in. × 11½ in. × 2½ in. (sometimes 3½ in.) thick. This was in 1700. He also gave descriptions, with illustrations of the bricklayer's tools; they were: brick trowel, brick axe, 'saw of tinn,' rubstone, square, bevel, a small 'trannel' of iron, float stone, little ruler, banker, pier, grinding stone, pair of line pins of iron, plumb rule, level, large square, ten feet rod, five feet rod, jointing rule, jointer of iron, compasses, hammer, rammer, crow of iron, pick-axe. To these he added the tools used in tiling and in plastering, and a list of general 'utensils,' viz. ladders, fir poles, putlogs, fir boards, 'chords,' sieves, 'lome hook,' beater, shovel, pick-axe, basket, hod, 'skreen' (and noted that with the latter one man could 'skreen' as much as two could with a sieve), boards, tubs.

From 1784 to 1850 bricks were taxed, and as this was per unit and not by weight, the consequence was that they were made larger. The tax is obsolete, and the size of bricks is now conditioned by the number which a bricklayer can lay as easily as possible in a given time. Until the third quarter of the nineteenth century, bricks were generally 2½ in. deep, at the present time they are usually 3 in. deep, and it is not likely that they will be made larger than this, as the thickness is fixed by the size of a man's hand, and bricklaying is not yet done by machinery.

Where stone was not available for dressings, the builders have generally had recourse to bricks for the mouldings. Now they are burnt to the required pattern, but in the year 1700 (according to Moxon), bricks for mouldings were formed roughly by the axe and then rubbed on a 'float stone' to the form desired.

Bricks are not necessarily burnt[2]; they may be only dried, and such unburnt bricks are in wide use at the present time. Throughout the countryside in Beaujolais and Mâconnais 'are small houses built of a species of sun dried bricks or lumps of clay...in times of flood those built in the river bottoms have been known to melt away like the sand castles of children at the seashore[3].' Mud bricks are also used in Albania. They are more easily made,

---

[1] *Dictionary of the Architectural Publication Society, s.v.* Brick Making.

[2] The traditional method of making burnt bricks has been described by Professor Beresford Pite, in the first *Building Construction* volume of the Architects' Library.

[3] Francis Miltoun, *Castles and Châteaux of Old Burgundy*, pp. 170 and 171.

and cost less than burnt bricks, and were used by the ancient Egyptians, by the Greeks and by the Romans, and their use by the English was therefore a following of the Roman tradition.

In Newcastle-upon-Tyne the fraternity of bricklayers was anciently styled 'catters and daubers.' Cats were pieces of 'straw and clay worked together into pretty large rolls and laid between the wooden posts in constructing mud walls,' according to the *Oxford English Dictionary*. Here we have evidence for the descent of the modern layers of burnt bricks, from earlier craftsmen who laid unburnt bricks. Brick nogging, such as is to be seen in the charming old cottages of Hertfordshire, may be regarded as a development, with burnt bricks, of the 'cat' form of construction.

The English took the use of sun dried bricks with them to America, and there they are used for filling in between studs, etc., and between inside and outside boarding.

In Norfolk, in the year 1847, cottages were built with clay walls and clay chimneys, the walls either in a solid piece or else with clay lumps or bricks well dried in the sun, and of any size which was thought convenient. They were carried up in a rather rough manner to ensure key, and the angles were protected by angle beads. The plaster might be either of good clay mixed with road sand or silt, or, more frequently, of old clay or loam and madgen, well kneaded with old straw to a proper consistency by being trodden by horses. Such a wall was raised 2 ft. above the surface of the soil on a flint wall, called a pinning. The buildings were cheap, and they were warm in winter and cool in summer[1]. Mr F. W. Troup says that 'Half the cottages in Norfolk that appear to be brick are really built of clay lumps faced afterwards (many quite recently) with $4\frac{1}{2}$ in. of brickwork[2].' According to another writer[3], these clay lumps of Norfolk are made of cob, pressed into moulds in the same manner as the clay for burnt bricks. The remaining processes are the same, but the lumps are used after drying instead of burning, and are of larger size than is usual with burnt bricks. The writer has found that many of the old cottages about Cambridge are built of clay lumps, which are unburnt bricks of large size. They are said to be

---

[1] *Builder*, v (1847), p. 388.
[2] 'Influence of Materials on Design in Woodwork' in *The Arts Connected with Building*, p. 68.
[3] Norman Jewson in a letter to *Country Life*, November 22, 1913.

durable and weatherproof when plastered externally —as they usually are—with a kind of chalk marl mixed with straw.

About the time of Waterloo (the year 1815) 'an eccentric character[1]' required a Suffolk bricklayer to build the walls of a barn of sun-dried bricks of clay and chopped straw, of a size of 2 ft. long, 1 ft. wide, and 9 in. thick. The eccentricity seems to have been not in the kind of bricks, but in their size, as the bricklayer found them very difficult to lift and lay. There is, therefore ample evidence for a use of dried bricks in England, and the descent of the Newcastle bricklayers, who use burnt bricks, from the catters, who used dried bricks, indicates, apparently, that the dried bricks, or 'cats,' or 'lumps,' were the earlier form. And the evidence from the Norfolk cottages mentioned above, seems to show that the sun-dried bricks were a development from the walls of mud or cob.

A prehistoric form of earthen or stone wall in which the materials were compacted by burning must be mentioned. The vitrification was done by the use of brushwood, and by lime where it could be obtained. An example near Bristol, is described in the *Antiquary*, 1911[2], and similarly constructed walls are also found in Scotland.

A peculiarly Northern form of wall is that made of turves, of which the most celebrated example is the Antonine wall, which was carried by the Romans across Scotland from Forth to Clyde. It is said that in it there may still be traced the dark lines caused by the decay of the grass, or other vegetation of the sods.

Allied to this are the walls composed of a mixture of turves and rough undressed stones, which latter might be laid in horizontal rows alternately with thin beds of turf, forming what was known as a 'spetchel' wall in Northumberland, and used for fences, or the wall might be of turves and faced on both sides or externally only with the stones, a form widespread in the Celtic parts of these islands. As an instance the walls of the black houses or 'clachans' of Lewis were of turf, with dry stone facings, five or six, or even seven feet thick. When it was wished to clean that part of the house in which the cattle lived, which was every Spring, the wall was pulled down and then built up again[3]. This kind of wall was used for the houses in the Isle of Man, but few now remain.

[1] *Builder*, v (1847), p. 364.
[2] James Baker, *A Neolithic Romano-British Settlement*, pp. 287 et seq.
[3] Sir Arthur Mitchell, *Past in the Present*, p. 51.

It is still very common everywhere in Wales in field fences, and to write of it brings memories of the purple heath and golden gorse and roughly cultivated fields of many a Celtic landscape. Dr Johnson in his *Journey to the Western Islands*, found walls in houses, which had the turf covered with wattlework to prevent it from falling. It is unsafe lightly to ascribe the origin of a form of construction to a particular race, for turf walls occur in Germany and in Scandinavia, and Mr Bernhard Olsen has informed me that they are to be found in Denmark, as are other kinds of English walls. There are, he says, in Denmark, walls of wattle and daub, walls of studs with the interstices filled with mud, walls of mud without either studs or wattle, the mud being mixed with straw, cow hair, cow dung, etc., and stone walls with the stones laid in mud mortar, all of which have their parallels in England. And as great a variety may be found in many districts of this country, thus, in the villages about Cambridge the writer has found house walls of (1) blocks of chalk or 'clunch,' banded with two or three layers of bricks[1], (2) ordinary bricks, (3) so-called 'home-made bricks,' $5\frac{1}{4}$ in. × $10\frac{1}{2}$ in. × 6 in. deep, (4) sun-dried bricks or clay lumps, (5) stud-work covered with ordinary lath-and-plaster, and also with split branches plastered, (6) wattlework, and (7) horizontal weather boarding[2].

The materials used in ordinary buildings in former times depended as much upon the ease with which they could be obtained, as upon the nationality of the builders. The Norman conquerors of England only imported their favourite Caen stone for buildings of the first importance, and when the Scandinavians settled in Iceland they had to abandon that lavish use of timber which was usual in the buildings of their home land.

---

[1] The original walls of Christ's College were similarly constructed. R. Willis and J. Willis Clark, *Architectural History of the University of Cambridge*, II, p. 222.

[2] The district is therefore East Anglian in its walling. See also note, p. 188.

CHAPTER XI

FLOORS

Earthen and mud floors—Saltpetre digging—The use of lime and sand for floors—Simple wooden floors—Joists and girders—The breast summer—The dragon beam—Projecting upper storeys—The ascent to the upper storey—The ladder and the stair—Floor boards—The old method of laying—Plastered ceilings.

The use of earth for the formation of floors was even more usual than it was for walls, for the most simple floor for a building is the ground on which the building has been erected. Mr H. W. Williams, writing of the huts, or 'clachans,' of Lewis, says, 'The floors of the huts are simply mother earth beaten hard by constant use[1].'

Such a floor would be repaired, when required, by the same material, 'mother earth,' and so there would be developed the earthen or mud floor, which was the most common kind of floor in ordinary buildings in older England. Such floors were in use everywhere, not only in the ordinary buildings but even in colleges and churches. The original floors of the 'Old Court' of Corpus Christi College, Cambridge, are said to have been of clay[2], and there is a record of the paving of the floor of Midhope Church, in South Yorkshire, in the year 1705[3]: earthen floors still remain in many of the old barns of the district.

Elementary as a mud floor seems to us, it had its technical excellence like more advanced work, and Henry Best, the East Riding farmer who wrote detailed instructions of the work to be done about a farm, and some of whose directions for buildings have already been quoted, also gave a description of his method of laying

[1] *Reliquary*, VI (1900), p. 76.
[2] R. Willis and J. Willis Clark, *Architectural History of the University of Cambridge*, I, p. 254.
[3] J. Eastwood, *History of Ecclesfield*, p. 483.

mud floors[1]. The earth was to be dug and raked until the moulds were 'indifferent small,' then water was to be brought in 'seas' and in great tubs, or hogsheads, on 'sleddes': the earth was then to be watered until it was as soft as mortar, or almost a puddle: it was then to be allowed to lie a fortnight until the water had settled and the material had begun to grow hard again and the floor then beaten down smooth with broad, flat pieces of wood. In mending holes in an earthen floor, red clay or clots were used, and rammed and beaten with old 'everinges and harrowe balls.' After the clots had been broken small with a mell, they were to be watered and left three or four days, for if they were to be rammed at once ('presently') they would cleave to the beater.

The generally used mud floors were porous, and so absorbed filth in the shape of nitrous matters. In order to search for and obtain the nitre required in the manufacture of gunpowder, an official generally known as the saltpetre man, had power to throw down the mud walls in common seisin, to take up the town hall floors, dig up the floors of stables, slaughter houses, etc. The account of the chamberlain of Leicester for the year 1584–5 contains an item for the making of the 'florthe ageyne in the kitchyn' at the hall after the saltpetre man, and in the year 1588–9 the chamberlain paid for the 'makinge agen of newe' of all the earthen floors at the hall after another visit from the destroyer[2].

The 'saltpetre man,' in his search for nitre, caused endless inconvenience and friction in the sixteenth and seventeenth centuries. He pulled up the floors of houses and cottages and there is a record of a saltpetre man even digging up the floor of a church in the year 1624. Ragdale Hall, in Leicestershire, fell into decay 'by the natural decay of its timbers and by the unwarranted depredations of saltpetre diggers' and was restored in the year 1633[3]. The diggers generally seem to have acted in a very high handed manner, and a book on the abuses and corruptions of offices spuriously attributed to Lord Coke, the jurist, states, 'This Salt-peter-man vnder shew of his authoritie, though being no more than is specified, will make plaine and simple people beleeue, that hee will without their leaue breake vp the floore of their dwelling house, vnlesse they will compound with him to the contrary. Any

[1] *Farming Book* (published by the Surtees Society), p. 107.
[2] *Records of the Borough of Leicester*, edited by Miss M. Bateson, III, pp. 221 and 255.
[3] *Architectural Review*, XXVIII (1910), p. 189.

such fellow, if you can meete with all, let his misdemenor be presented, that he may be taught better to vnderstand his office : For
by their abuse the country is oftentimes troubled[1].' The requirements seem to have become more onerous in the reign of King
Charles I, and the abuse of saltpetre digging was swept away
during the Commonwealth.

Mud floors were very dusty and difficult to keep clean, and it
was no doubt the dust of a mud floor which Puck had to sweep
behind the door in *A Midsummer Night's Dream*[2]. The universal
dustiness of mud floors, noticed by Sir Frederick Treves[3], even in
the huts of Central Africa, must have led to attempts to make the
mud floors less dusty. A simple method was to strew the floor
with rushes or other green and therefore damp plants. In another
method borrowed from Italy at the time of the Renaissance, the
mud for the floor was mixed with a certain proportion of bullocks'
blood, which made it hard and gave to it the appearance of black
marble when polished. Sir Hugh Platt, writing in the *Jewel House
of Art and Nature*, in the year 1594, recommended for ground
floors a composition of fine clay, mixed with ox blood, as making
a smooth, glistening and hard floor, and the composition would also
serve as plaster for a wall[4].

Bullocks' blood was also used as a substitute for paint on woodwork : it was mixed with soot or 'rud.' 'So late as 1861 the
south door of York Minster was treated with a composition of rud
and bullocks' blood[5].'

Another method of preventing the mud floors from wear was to
mix with them bones, which presented a resisting surface, and Dean
Aldrich, an amateur architect, and a contemporary of Wren, wrote
in his dull book, the *Elements of Civil Architecture*, that in his
day it was quite usual to do so. In the excavations at Chesters,
the station Cilurnum, on the Roman Wall in Northumberland,
quantities of bones were found in the floors, and Dr Collingwood
Bruce took this to be an example of the deterioration of discipline
among the Roman soldiers in the later days of the Roman occupation[6] : but it seems possible that the bones were the remains of
floors in which the other materials have perished.

[1] *Notes and Queries*, 1st Series, VII, pp. 376, 433, 460, 530.
[2] Act V, Scene II.   [3] *Uganda for a Holiday*, p. 208.
[4] Receipt No. 90, page 72, in the edition of 1653.
[5] *Notes and Queries*, 9th Series, XI, p. 499.
[6] *Handbook to the Roman Wall*, 5th edition, p. 100.

Bones were often driven into the earthen floor, to form a pattern, in the seventeenth century, and an example in Broad Street, Oxford, destroyed in 1869, 'was laid with "trotter-bones" in a pattern of squares, arranged angle-wise within a border. The pattern was defined by bones about 2 in. square, rubbed or sawn to an even surface, and filled in with the small bones of sheep's legs, the knuckles uppermost, closely packed and driven into the ground to the depth of from 3 in. to 4 in.[1]'

An improvement took place when lime, a material with valuable setting properties, was mixed with the floor material. Lime may have been used at first as a substitute for clay in districts where clay was unobtainable, as in certain parts of North-East Derbyshire. Although floors containing lime seem to have been in use from the Norman period, they were only put into ordinary buildings at the time of those improvements in English agriculture which accompanied the Napoleonic wars, the growth of the population, and the inclosures of the common lands. The use of lime and sand for the floors of cottages was new to William Marshall when he wrote his *Rural Economy of Yorkshire*, in the year 1787, and he found that they were mixed together as for mortar in bricklaying, except that the mixture was made stronger and softer[2]. In Gloucestershire, in the year 1807, Thomas Rudge found a floor in fashion called a 'grip floor,' composed of lime and ashes, or other materials, laid in a moist state to the thickness of 4 in. or 5 in., and worked or rammed with a heavy slab of wood, till it acquired hardness and a smooth surface: it was used on ground floors, but he did not think it suitable for parlours, cheese stores, etc.[3] In the Midlands to-day such floors are known as 'lime ash floors,' and the many examples which are still in existence in the upper stories of houses of the seventeenth century show that they were then very fashionable among the smaller gentry. For a time they superseded boarded floors.

In Derbyshire they were usually laid upon laths, and farther to the south upon straw. A piece of a floor from Leicestershire, which was presented to the writer, was laid upon rather poor wheat straw, with the threshed ears still attached. In North Derbyshire the writer has heard the name 'plaster floor,' and the same term is

---

[1] *Building News*, September 3rd, 1869.
[2] I, p. 135.
[3] *General View of the Agriculture of the County of Gloucester*, pp. 45–6.

used in the Midlands. Mr F. W. Troup has described the floors of this kind in a house in Norfolk, which he repaired: 'The floors were solid, and were formed in this way: joists 4 inches by $3\frac{1}{2}$ inches were laid flatwise on the main beams and were covered with a layer of reeds about an inch in thickness: on this again were laid wide oak floor boards nailed to the joists through the reeds. This being done, the underside of the reeds was plastered between the joists, giving a good, sound proof floor and yet barely 3 inches thick. In Leicestershire and other districts, where good lime is to be had, many of the old floors were formed altogether of plaster. In place of the floor boards laid on the layer of reeds, 2 ins. or 3 ins. well tempered plaster is spread and thoroughly floated over and brought to a smooth surface. This gets almost as hard as cement concrete, and the under surface of the reeds is plastered to form the ceiling as before[1].' Mr Troup continues with enthusiasm ' Here is a good solid floor three times as sound proof as ferro-concrete, all construction showing, not more than 4 inches thick, has been done for centuries, and much more sound proof than steel and concrete.' Evidence of Elizabethan date for the use of the lime floors is contained in a letter from Elizabeth, Countess of Shrewsbury, generally known as ' Bess of Hardwick,' who wrote various instructions from London to her servant Francis Whitfield, at Chatsworth, such as that a good store of beer was to be prepared, and that the floor of her bed-chamber was to be made even either with plaster, clay, or lime.

In expensive buildings the marble paving of the Continent was imitated by stone 'flags' and these descended in the course of time, as usual, to the ordinary buildings in such districts as South Yorkshire, where similar stones could be easily obtained. Mr Ralph Nevill found two cottages 'near Black Heath on what was till recently the Grantley estate' in one of which one-half of the ground floor 'retained what was probably its original flooring of rough slabs of Horsham stone, fitted together like a puzzle. Brick paving as in the other half is always of a much later date. Where stone could not be conveniently procured, beaten earth was used covered with rushes[2].'

Timber has now taken the place of earth or mud as the usual

---

[1] 'Influence of Materials on Design in Woodwork' in *The Arts Connected with Building*, p. 70.
[2] *Old Cottage and Domestic Architecture in South-West Surrey*, p. 12.

material for the floors in ordinary buildings : like them its use
reaches back to Romano-British times, and from Anglo-Saxon
times instructions for a reeve ('gerefa') have come down which
state that among other duties he was to bridge between the houses
('betweox husan bricgian'). Dr C. M. Andrews thought that the
bridging was of rough logs, possibly hewn on one side, to protect
from the mud of Winter and Spring, which would have made a
very simple form of wooden floor[1]. At the destroyed Gatacre
Hall, described in an earlier chapter, there was a somewhat more
advanced kind of floor, which was 'of oak boards three inches
thick, not sawed, but plainly chipped[2].' The makers of these
floors had apparently no knowledge of the need for a circulation
of air round the wood, and they differed from our modern floors
because there was no distinction between the floor proper and its
supports.

In the framed structure found in Drumkelin Bog, previously
mentioned, the upper floor was formed in the same manner by split
trees laid close together. The splitting indicates some advance in
culture, for in a simpler form the trees would have been left
unsquared. In some existing examples the influence of the usual
mud floors of the ground storey is shown by a covering of mud
laid over the poles or logs. Mr T. Winder has illustrated such a
one from the millyard at Derwent, Derbyshire: where 'larch
poles are laid side by side close together, and a mud floor is
plastered over them[3].' Moritz Heyne, in his *Deutsche Wohnungswesen* also, has cited a very early German example.

According to Mr John Ward, the use of timber was exceptional
for the ground floors of Romano-British buildings. At Silchester,
a patch of gravel in Insula XVI had a series of parallel trenches
about 6 in. wide and 18 in. apart, which were assumed to be the
spaces in which joists had rested. Floors of beaten earth, gravel
or clay were mostly confined to outbuildings and cottages and
rooms of little importance. Mortar and concrete floors were
frequent and were finished with the finest white lime. Flagged
paving and pitching were chiefly used for stables and outbuildings
and especially for yards and other spaces open to the sky[4].

Until comparatively recent times in England, simple buildings

---

[1] *Old English Manor*, p. 205.   [2] Ante p. 117 and *Archaeologia*, III, p. 112.
[3] *Builder's Journal*, III, p. 41 (February 25, 1896).
[4] *Romano-British Buildings and Earthworks*, pp. 265 et seq.

of one storey, with vertical walls, often had flat ceilings, which were used as the floors of lofts, and when they were formed of boards laid side by side, as was done sometimes for the strength, and perhaps also for the sake of appearance, the boards were laid alternately 'thick and thin.' This method was also in use on the continent; a Swiss example of a floor formed in this manner was illustrated by Gladbach in his *Schweizer Holzstyl*. From this early boarding a method of forming the upper floors in old South Yorkshire buildings was developed in which every alternate board was thicker, and was rebated on each edge above and chamfered on each edge underneath, and the alternate thinner boards were laid in the rebates, and in this manner a level floor was formed. In Fig. 52 such a floor is shown in section. The

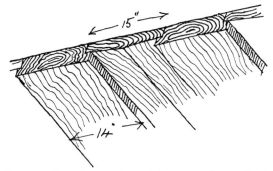

Fig. 52. Floor of upper storey at Dove Hill Houses, Sheffield (now destroyed). There were no joists and the floor boards were laid thick and thin alternately. The method was probably copied from boarded ceilings.

writer found it at Dove Hill Houses, Endcliffe, Sheffield, when they were being demolished, and it will be seen that the thinner boards were $1\frac{1}{2}$ in. by 15 in., and the thicker were $4\frac{1}{2}$ in. by 14 in.

Samlesbury Hall provides us with a floor of a rather more advanced type; here according to Whitaker, 'the boards of the upper floors, which are indeed massy planks, instead of crossing, lie parallel to the joists, as if disdaining to be indebted to the other for support[1].' This form of flooring is more suited to ground floors and may have been copied from wooden floors laid directly on the ground, like the floor at Silchester before mentioned.

Our modern 'board-and-joist' floors are an improvement on the type, because the boards are laid across and not with the joists, and

[1] *History of Whalley*, p. 500.

so the weight is distributed.   Mr Winder found a practical appre-
ciation of this in a floor near Sheffield, in which the deeper boards
of a floor like that at Dove Hill Houses had been reused as
ordinary joists, and the thinner boards had been reused to form
cross-laid floor boards of the usual kind.   The earliest existing
examples of ' board-and-joist' floors in England are in the Anglo-
Norman castles.   Mr G. T. Clark wrote that 'the floors of these
keeps are almost always of timber : thick rough planks, resting
upon stout baulks 12 in. or 14 in. square, placed about 2 ft. or 3 ft.
apart, and resting either on a ledge or in regular joist-holes[1].'

When the ' bearing ' of the joists between the walls was too long
to carry without support, beams at right angles to the joists were
used at intervals, and from this was developed the elaborate
' framed floor ' of the nineteenth century books of carpentry and
joinery.

The cross beams were known as girders, summers or somers, and
dormants : one of them carried the chimney, and so was called the
' bressummer,' that is, the breast girder.   It ran at some distance in
front of the fireplace, and was generally boldly chamfered on the
lower edge on the side towards the fire, and two short transverse
girders were run into it to carry the sides of the chimney.   Such
beams were in general use in houses in the Sheffield district, until
they were superseded by the less inflammable fireplace arch about
the middle of the seventeenth century (see Fig. 72).

The early builders knew nothing of statics and of formulae for
determining the strengths of bearing timbers, and so they used square
joists, as in the Norman castles, or if the joists were not square they
laid them with their broad sides horizontal : probably this was due
to the original use of whole or split undressed logs, as was the case
with the rafters.   Often in the mediæval period, the space between
the joists hardly exceeded the width of the joists.

Moxon wrote that it was the custom in his time to lay the
bressummers and girders flat upon their broadest sides, ' but the
Joysts are to be laid contrary' by reason of their mortices[2].

In houses of the fifteenth or early sixteenth century, with a
projecting upper storey, which were common in the south-east of
England, but rare in the north, a large beam was run diagonally
across the building, and into this the joists of the upper floor were

---

[1] *Medieval Military Architecture in England*, 1, p. 126.
[2] *Mechanick Exercises*, House Carpentry, p. 139.

framed : the joists on each side of the beam were thus at right angles to each other. The beam is called a dragon beam, and it has been assumed that this is an ignorant corruption of diagonal beam: more probably it is the beam upon which the joists seem to draw or pull. When the ground floor walls were of timber, the ends of the dragon beam were carried by an angle or corner post, which was placed upside down, as was usual with posts in the Middle Ages, and the butt of the tree forming the post was treated as a corbel. It has been stated that this was for the purpose of allowing the sap to run out by the same way as it had entered the timber, but it is uncertain whether this is traditional. The angle post was fixed with the projection either to the interior, or to the exterior; in the latter case it was visible from outside, and was often carved, and examples remain in many old English towns. The use of the diagonal beam arose from the practice of forming the projecting stories on all sides of the house. Various reasons have been advanced in explanation of the projecting upper storey, which was so ubiquitous in the later Middle Ages, such as the desire to gain additional space in the upper storeys without encroaching upon the inconveniently narrow street : but projecting upper storeys are found in country houses, where economy of space was not important. It has also been supposed that it was done to protect the timbers of the ground floor from the weather : but if so the projecting upper portions received no benefit. The earliest timber buildings in Western Europe, with upper storeys other than lofts in the roofs, seem to have been towers in connection with defence, and these with their projecting 'hourdes' all round probably served as the models when dwellings were built with an upper storey. In the Middle Ages such a projection of an upper storey was called a jetty or jutty[1].

After 'board-and-joist' floors had been found convenient on the upper floors, they were used on the ground floor: but their use is rare in that position, in ordinary buildings, before the nineteenth century. Mr E. Guy Dawber has written that the upper floors of the old cottages and farmhouses of the Cotswold district 'were generally of unsquared joists of timber, often with the bark left on, laid some few inches apart, and packed in between with a

---

[1] Wm Horman in his *Vulgaria* (1519), p. 246, translates ' Buyldynge chargydde with jotyes ' by ' Mœniana Aedificia,' which implies that the projecting upper storey was for defence.

mixture of clay and chopped straw on interlacing hazel sticks. Underneath the ceiling was plastered, with the floor above either laid with oak boards or else finished with a smooth cement face[1].' The ground floor was 'of large slabs of stone, laid directly on the earth....This method of laying floors was but a slight advance on that of some century earlier, when the natural earth well trampled upon formed the only flooring[2].'

Old floor boards were variously sawn, axe dressed, or riven ; the latter are the stronger, but when of oak, which was the usual material, they are dimpled, and unless the grain of the tree from which they are split is straight they cannot be used in long lengths. In an old mill at Hathersage, Derbyshire, the riven oak floor boards were laid in lengths of about 3 ft. only. In the year 1507–8 the churchwardens' accounts of St Michael's, Bath, show that a tree was bought, and then the wardens paid for its squaring and carriage, and seven shillings for sawing it into seven hundred boards[3]. Sometimes the tree was quartered with the result that the boards of the floor were of varying widths. In old floors the boards were pegged to the joists, and there were various methods of making the joint between joist and girder.

Moxon described the method adopted in 1677 in laying floor boards[4]. In the first place the boards were rough planed and seasoned. When the first board was laid, a second board was laid close to it. ' But before they nail it down they again try how its sides agrees with the side of the first, and also how its thickness agrees with the first Board. If any part of its edge lie hollow off the edge of the first Board, they shoot off so much of the length of the Board from that hollowness towards either end, till it comply and make a close Joint with the first. But if the edge swell in any place, they plane of that swelling till it comply as aforesaid. If the second Board prove thicker than the first, then with the Adz (as aforesaid) they hew away the under side of that Board (most commonly cross the Grain, lest with the Grain the edge of the Adz should slip too deep into the Board) in every part of it that shall bare upon a Joyst, and so sink it to a flat superficies to comply with the first

---

[1] *Old Cottages, Farmhouses, etc. in the Cotswold District*, p. 15.

[2] Ibid. p. 18.

[3] *Proceedings of the Somersetshire Archæological and Natural History Society*, 1877–80.

[4] In the ' House Carpentry' section of his *Mechanick Exercises*, pp. 155 et seq

Board. If the Board be too thin, they underlay that Board upon every Joyst, with a Chip.' 'And as this second Board is laid, so are the other Boards laid.' When the first and fourth had been nailed down the second and third were set at an angle and two or three men jumped on them or else they were forced down with 'Forcing Pins and Wedges.' Thus the boards were laid in groups of three, and every third board was nailed down first. Two brads were used to every joist, each fixed 'about an Inch or an Inch and a half within the edge of the Board.'

At first the boards and joists of the upper floors were not concealed, but showed in the room below : such ceilings were dark and in the course of time were thought to be unsightly. For these reasons ceilings of plaster on laths were introduced from the Continent, and became usual in great houses in the sixteenth century. Then, as was wont, their use gradually filtered down through the strata of society.

The old plastering laths were hand riven with a two handed shave, and though very rough in appearance, they are stronger than sawn laths of similar material and section of the present time, for in the riven laths the grain of the laths was not cut through.

The plastering of ceilings to hide the woodwork of floors or of roofs was called 'under drawing' : and plastered ceilings, like the plastered walls and partitions, were sometimes filled with chaff. In East Anglia, in upper rooms open to the slope of the thatched roof, the thatch itself was plastered between the rafters, and this was known as 'sparkling.' In South Yorkshire, in the seventeenth and eighteenth centuries, thatching materials were used to form plastered ceilings, either sloping on the undersides of roofs of stone slates, or flat on the undersides of wooden joists. This was done by securing layers of thick-stemmed reed-grass—as true reed was not obtainable—to the rafters or joists by laths nailed over the grass, parallel to and on the undersides of the rafters or joists. In a Sheffield example, of the eighteenth century, there were, in addition, occasional cross laths, fixed in and parallel to the reeds no doubt in order to stiffen them. In an old building at Worksop, Nottinghamshire, the writer found this method used for the outer walls, wheat straw being laid across the studs and secured with laths. The accounts of the Corpus Christi Gild of Leicester show that in the year 1493–4 'strey laths' were bought for repairs to a house[1] : these were used to fasten the straw or reeds to the joists or beams.

With the knowledge that joists were stronger in proportion to their depth than to their thickness, came a use of thinner and deeper joists, which had a tendency to buckle, and it was found that a saving in cost could be effected by the use of cross stiffeners with the thin joists. At first thin pieces of joist were used for the stiffeners but later herring bone strutting was devised, and this has almost superseded the simpler form.

For long periods ordinary buildings were only of one storey, and in the year 1254, according to Mr T. Hudson Turner[2], the houses of London seldom, if ever, exceeded two storeys, including the basement. So, in Suckling's *Ballad upon a Wedding*, written about three hundred years ago, a countryman is made to describe a nobleman's London mansion as 'a house with stairs,' which was supposed to be a curiosity to the countryman. After upper storeys had become usual in buildings in the fifteenth century, they seem sometimes to have been added to existing houses. The accounts of the churchwardens of St Michael's, Bath, for the year 1415, contain an item of twenty pence for making a solar at a house then occupied by John Peres[3]. Such upper storeys required some means of ascent, which was provided either by ladders, or by stairs.

The access to the upper storey of buildings in early times was either on the outside, or inside, and the order of its development was probably as follows :

(1) An upright pole with pegs driven through it and projecting, for the hands and feet.

(2) A ladder, a method which was in use in houses in South Yorkshire as recently as the last century: it is the 'stee' of the Northern dialects, and the 'staeg' of Aelfric's translation of 'ascensorium' : in the fifteenth century *Catholicon Anglicum* there is 'a laddyr, ubi a stee.'

(3) A staircase cut out of a solid baulk of oak in a straight flight: an example at Rake House, Surrey, has been noticed by Mr Ralph Nevill[4].

---

[1] M. Bateson, *Records of the Borough of Leicester*, II, p. 346.

[2] *Domestic Architecture in England from the Conquest to the end of the Thirteenth Century*, p. 95.

[3] In 1415. *Proceedings of the Somersetshire Archæological and Natural History Society*, 1877–80.

[4] *Old Cottage and Domestic Architecture in South-West Surrey*, p. 11.

(4) A staircase in a straight flight as the last, but each step is cut separately out of the solid timber[1]. This form was copied in stone, and more recently in concrete, and therefore is still in use.

(5) A combination of the first and second types, the so-called circular newel steps, in which the newel is a straight, upright tree and the steps are framed into it, is shown in Fig. 53, from a ruined house of the early sixteenth century at Oxspring, South Yorkshire.

Fig. 53. Circular oak staircase in the ruins of Oxspring Lodge, South Yorkshire. It is said to have been made from a single oak tree. The newel is in one piece from top to bottom. The undersides of the solid steps are dressed to form one smooth surface on the underside.

Tradition says that the whole staircase was made from the timber of one oak tree.   There is a similar staircase, with a very long newel, at Plas Mawr, Conway, and the form is common in buildings of the fifteenth and sixteenth centuries.   The circular newel stairs in stone of the mediæval churches are, of course, copied from those in wood ; but the appearance of the upright newel tree was only

[1] An example has been illustrated by Mr Addy in *The Evolution of the English House*, p. 57.

obtained by working a portion of the newel on each step[1]. The stone newel staircase, as we know it to-day, must be the perfected result of attempts which have not come down to us, for when one craft copied the work of another, it was some time before the old design was adjusted to the new material. The old craftsmen only attained to the correct adaptation of design to material after gradual experiment: although the right use of material was usual with them, it was the result of previous failure, and not of instinct.

(6) The modern framed staircase, with separate strings, treads, and risers.

'The modern form,' says Mr Ralph Nevill, 'in humble buildings, at least, is a sure sign of a date later than 1600[2].'

There are two varieties of ordinary stairs, the one with close and the other with cut strings. The former is the earlier, and the latter superseded it in the eighteenth century, but since the nineteenth century both kinds have been in use.

[1] Thick trunks of trees were probably used for the old wooden newels, as the newels of the stone stair in the tower of Hough-on-the-Hill Church, Lincolnshire, are 18 in. diameter in the portion supposed to be pre-Conquest, and only 9 in. in the later mediæval portion above. W. F. Rawnsley, *Highways and Byeways in Lincolnshire*, p. 71.

[2] *Old Cottage and Domestic Architecture in South-West Surrey*, p. 12.

CHAPTER XII

SLATED ROOFS

Stone slates—Their wide use in England—True slates—Their comparative advantages—Fixing of old Welsh slates—Iron nails and wooden pins—Old names of slates—Bedding slates—Mossing—Ridges—Shingles and slats—Plain tiles—Pantiles.

The development of the main framework of the building, of the walls and the floors, has been described in the preceding chapters. In this chapter and that which follows, the old methods of covering the roof are described, as far as that is possible, for the roof covering is the part of a building which most readily decays, and it is doubtful whether there are any roofing materials which are old, in the historical sense, in position to-day, with the possible exception of some lead. But lead, like other metals, was formerly too costly for use as an ordinary building material and therefore has been placed with the newer materials in chapter XVI.

The old English names for the covering of buildings were 'roof' and 'thack.' The latter, still in use in the dialects, has been softened into 'thatch' in modern literary English, in accordance with custom, and is now only applied to roof coverings of vegetable matter, such as straw, reeds, and heather, but the word was originally applied to roof coverings of any kind. In the thirteenth century there was a church in Lincoln called St Peter Steintheked, that is, stone roofed, and in the fifteenth century building accounts of York Minster, the boards, upon which the lead was laid, are called 'thackburdes,' which means roof boards[1]. In more modern times, on October 1st, 1647, Adam Eyre, a South Yorkshire Puritan captain, entered in his diary[2] that he had been 'to James Mitchell, who promised mee to come to mend the thatch to-morrow,' and on the following day Mitchell came with five men, 'and putt up other

---

[1] *Fabric Rolls of York Minster*, pp. 39, etc. [2] p. 65.

stones on the wayne house, and I payd them at night for their worke and for 2 loads of slate, and nayles, 7s. 4d.' The slates were the usual stone, or 'grey,' slates of the district, and Eyre used the word 'thatch' in its older and wider sense of roof covering. The word is still in use with the older meaning in the West Riding, where children playing at ball will say 'my ball's gone on t'thack,' the 'thack' now being usually Welsh slates. It is evident that thatch meant any roof covering, and the present restriction of the word to roofings of vegetable matter shows that in the past their use was general, and that of other materials exceptional.

In the south and west of England roof covering was also called 'heleing.' In *Pierce the Ploughman's Creed* there occurs :

> Then came I to that cloystre and gaped abouten,
> Whough it was pilered and peynt and pertreyed wel clene :
> Al yhyled with lead, lowe to the stones,
> And ypaved with poynt tyl ich point after other.

Here the word 'hele' is applied to lead, and the spirit of the poem shows that the leaded roof and the floor of diagonally laid tiles must have seemed unusual and ostentatious to its fourteenth century writer.

At the present time the cheapness, durability, and convenience of Welsh slates have led to their almost general use as the covering of ordinary English buildings.

The restriction of the word 'slate' to rocks which possess 'slaty cleavage' is recent and due to the geologist, but the older builders used the word for any stone which could be or was naturally split into flat, tile-like layers, and such 'slates,' made from fissile sand-stones or limestones, were in general use in many districts of England from an early period. They are common in the English Pennines, and in the Sheffield district they are known as 'grey slates,' to distinguish them from the Welsh, or 'blue slates[1].' In Scotland, in the year 1808, according to Jamieson, the stone slates were called 'brown slates[2].'

The best known of the fissile stone 'slates' are those of Horsham, which were used in Kent, Sussex and Surrey. In the south-west part of the latter county, according to Mr Ralph Nevill, they were superseded by tiles[3]. Other well-known stone

---

[1] Such 'grey' slates are shown in Figs. 15, 43 and 63.

[2] *Etymological Dictionary of the Scottish Language*, s.v. Skaillie.

[3] *Old Cottage and Domestic Architecture in South-West Surrey*, p. 32.

slates are those of Colley Weston, and Easton, in Northampton-shire and Rutland. The latter were used on the large barn built at Peterborough in the year 1307, and destroyed some twenty years ago.

In the year 1530 Palsgrave wrote 'I sclate a house with stone-slates: it is better to sclate a house than to tyle it.' Stone slates were appreciated and used in many districts. William Stephenson recorded, in the year 1812, that stone slates were used in the Isle of Portland and round Sherborne[1], and at the same time Thomas Rudge, another agricultural topographer, found that 'stone tiles' from the Cotswolds were used in Gloucestershire, or raised in the neighbourhood, as in the Forest or in the lower part of the Vale[2].

In the year 1813 Archdeacon Plymley noted that there were some good quarries of stone slate in the south-western district of Shropshire[3]. Mr W. F. Price, writing quite recently, in an article on the homes of the Lancashire yeomen and peasantry, says that many houses in the district of Tunley are roofed with Upholland flags, fastened on with oak or elm pins[4].

Professor Thorold Rogers has recorded the use of the fissile stones of Stonesfield as roofing slates at Oxford in the Middle Ages, and he found that 'slates' were bought in the Isle of Wight, Hampshire, Sussex, Kent, Wiltshire, Somerset, and Oxfordshire, and from Guiting in Gloucestershire. He thought that the slates procured at Southampton, the Devon villages, and Battle were possibly French slates from Angers[5]. In the fifteenth century the churchwardens of St Michael's, Bath, bought 'tyle stones,' which are said to have been of Bath stone from quarries on Lansdowne[6]. Steetley Chapel in Derbyshire, a building of the twelfth century and long roofless, was found at the 'restoration' in the last century to have been originally covered with stone slates, which were probably quarried from the magnesian limestone in the parish.

The use of fissile stones for roofing was widespread in England during the period of the Roman occupation. General Pitt-Rivers,

[1] *General View of the Agriculture in the County of Dorset*, p. 85.

[2] *General View of the Agriculture in the County of Gloucester*, p. 46. Many illustrations of buildings in the Cotswolds, covered with stone slates, will be found in Mr Guy Dawber's *Old Cottages, Farmhouses, etc. in the Cotswold District*.

[3] *General View of the Agriculture in the County of Shropshire*, p. 106.

[4] *Memorials of Old Lancashire*, I, p. 250.

[5] *History of Agriculture and Prices*, IV, p. 442.

[6] *Proceedings of the Somersetshire Archæological and Natural History Society*, 1877–80.

in his excavations of Romano-British villages, found roofing tiles or slates of Purbeck shale, with a hole in the middle of the head, as in the early Welsh slates, and in one of them was an iron nail. At Woodchester the tiles were of the gritty stone found near Bristol or the Forest of Dean[1].

The above evidence, so varied in date, shows that 'slates' have been in use over wide districts of England from the beginning of the historic period. In some of the districts the use of these stone 'slates' or 'tiles' was restricted to the most important buildings. Excavations in Roman stations, such as Gellygaer, have shown that some buildings were tiled and others were covered with materials which have perished, probably thatch. And much later, at Ripon, in the year 1399, while the roof of one tenement was being thatched with straw, one William Sclater was selling to the Chapter a thousand slate stones for the roof of another tenement[2]. In one district of South Yorkshire 'grey slates' have been in general use for so long a time that no tradition remains of an earlier form of roofing ; there is a 'sclaster' in the Poll-tax returns of the district for the year 1379. There was a family of the name 'Stonethacker' at Chesterfield in the fourteenth and fifteenth centuries. The earliest record known to Whitaker, of a slated house in the Craven district of Yorkshire, was Appletreewick Hall which was rebuilt about the beginning of the fourteenth century. Whitaker says that, at the same time, the choir of Long Preston Church was covered with shingles[3]. In the Midlands, the *Records of the Borough of Leicester* for the year 1377–8 show that 6s. 3d. was paid for 2500 slates, then occurred the cost of carriage, and further items for dressing ('baterer'), and for holing ('percer').

Sir Balthazar Gerbier, giving *Council and Advice to all Builders*, in the year 1662, considered that lead and 'blew slates' were the best roof covering for a house, and Holme[4] in 1688 thought that slates of 'blew slaggy Marble,' were the best for duration considering the cheapness, 'though the last in time,' but the wide use of true slates is comparatively recent outside of the districts in Wales and the English Lakes in which they are now quarried.

---

[1] Further information as to Romano-British slates will be found in Mr John Ward's *Romano-British Buildings and Earthworks*.

[2] *Memorials of Ripon*, published by the Surtees Society, III, p. 130.

[3] *History of Craven*, 3rd edition, p. 512.

[4] *Academy of Armory and Blazon*, p. 266.

The Welsh and the so-called Westmorland slates were but little used beyond the districts in which they were obtained before the introduction of transit by canals, and the resulting decrease in the cost of carriage, at the beginning of the nineteenth century. Since then the extension of the use of true slates has been coterminous with that of the railways.

In the first quarter of the nineteenth century the grey slates were being ousted in Sheffield by the importations from Wales and Westmorland, and the excellent qualities of the imported slates are set forth in an advertisement by a firm of slate merchants, in the *Sheffield Courant*, August 10th, 1827. The advertisement says: 'The advantages of Blue over Grey slating are many and important, being more durable and so light that 24 cwt. will cover 60 yards, the grey requiring six tons. A great saving of Timber is also effected, making a difference of 5*d*. per square yard, after allowing a good substantial roof: besides, when plastered inside, they make excellent shop ceilings: and being nailed to the laths with copper nails, on each side, they are made so firm as to render nugatory the most boisterous attacks of the weather: whereas the ponderous weight of the Grey Slates so bulge and otherwise damage the walls of the Houses, that they have frequently to be taken down, or secured with Cramp Irons, thereby incurring danger, increasing expense, and lessening the appearance and value of property: besides, they are early subject to moulder and decay, and destroy Ceilings, Goods etc., and not unfrequently by falling, endanger the safety of the inhabitants. Many Grey Slate Roofs, which appear good outside, are nevertheless decayed: and were it not for the roofs being flat would drop into the street.' The advertisers stated the case against stone slates in its blackest form.

Mr Guy Dawber has put the case of the Cotswold slates more fairly: he says that they are dearer and heavier and so require stronger roof timbers, and need more care and trouble in fixing, and they also need occasional repairing, but if the slates are properly seasoned at first they are almost imperishable, for no frost or wet will touch them. Houses and barns two hundred and more years old still stand covered with their original roofs[1].

Stone slates are now, owing to their cost, more or less luxuries; the small quarries, or 'delfs,' are closed from which the old slates

---

[1] *Old Cottages, Farmhouses, etc. in the Cotswold District*, p. 51.

for the adjoining farm houses were obtained, and the present stone slates are brought by railway from large quarries which specialise in their production. The use of stone slates is being revived, but it is anachronistic and also uneconomic, like the revival of handi-crafts, of which the products cannot compete economically with machine-made goods. Nearly everywhere in England the Welsh slates are the economically sound material for roofing and this has its significance for the Art of Architecture, if the old rule, that

Fig. 54. Old building at Criccieth, Carnarvonshire. The slates of the roof are small and thick, and the walls are built of large rough blocks.

the materials most æsthetically suitable for a district are those which are obtained in that district, is economically unsound. To that extent a rule of Art is no longer natural but exotic.

In the year 1813 in Shropshire Archdeacon Plymley considered that 'at the present price of straw, the comparative expense of blue slates which are gotten from the neighbouring counties of Wales is not excessive[1].'

Mr W. G. Collingwood, writing of High Furness, says that little slate was exported earlier than the eighteenth century, but it was locally used for roofing long before that time. Many of the

---

[1] *General View of the Agriculture of Shropshire*, p. 106.

seventeenth and eighteenth century buildings of the district were thatched, but from the ruins of old houses which have been excavated and dated it appears that slating was not uncommon, though before the eighteenth century the 'riving,' or splitting of the metal, had not reached its modern standard[1]. The early users were content with slates from the outcrop, and did not quarry for them.

The oldest Welsh slates that the writer has seen on buildings in North Wales are thick and small, like the smallest sized Westmorlands used towards the ridge in random slating. The illustration (Fig. 54) shows this kind of slating on an old building at Criccieth, Carnarvonshire, and it also shows a local form of old rubble walling. The slates were fixed with oaken pegs. In the *Old Cottages of Snowdonia*, Messrs Hughes and North state that they have found that the earliest slates in that district have an average size of 5 in. by 10 in. They had a peg hole in the middle of the head, that is, the top of the slate, by which they were hooked to the wattling[2]. The writer finds that the little old Welsh slates used on some of the Cambridge colleges are known locally as 'rag slates.'

In the year 1688, according to R. Holme, the slater's tools were as follows: Hatchet, Trowel, Hewing Knife, Pick to Hole, Pinning Iron to widen the holes, Hewing Block, Lathing Measure, and Stone Do., and Pins, Stone Nails or Lath Nails and Lath or Latts[3].

Romano-British slates were of various shapes. At the station of Caersws, and at the villa at Bisley, Gloucestershire, they were hexagonal, which was a usual form : at Mitcheldever Wood, in Hampshire, they were oval-shaped, and this form continued in use until the Middle Ages. In later times the 'tail' of the slates was cut square, and the head left a rough oval, as Westmorlands are at the present day.

Professor Thorold Rogers found that the fissile stone of Stonesfield, Oxfordshire, in the Middle Ages was made into three sizes of slate, viz. (1) common large, (2) middling, (3) large[4]. He thought that slate making in the Oxfordshire quarries was a

---

[1] *Memorials of Old Lancashire*, II, p. 179.
[2] The authors also describe the later developments of slating in North Wales, in their book.
[3] The tools are described in the *Academy of Armory and Blazon*, p. 395.
[4] *History of Agriculture and Prices*, I, p. 423 and IV, p. 442.

by-industry, for building stones were obtained from the quarries as well as slates, and he supposed that the quarrymen, when the demand for building stone was slack, employed their time in splitting, dressing, and boring such stone as was available for slates.

The Welsh slates at the present time can be procured in a number of stock sizes, which are named after the titles of ladies of the aristocracy; these names are quite recent, having been adopted about the year 1750; they are said to have been given by General Warburton, the proprietor of some large quarries in North Wales. Mr Guy Dawber has shown that the Cotswold slaters also distinguish the various sizes of the local stone slates by different names, which are not the same as those of the Welsh slates. The Cotswold names, although quaint, do not seem to bear evidence of any great antiquity[1].

R. Holme, in his *Academy of Armory* published in the year 1688, gave the contemporary names of slates 'according to their Several Lengths' as follows: Short Haghattee, Long Haghattee, Farwells, Chilts, Warnetts, Shorts, Shorts save one or Short so won, Short Backs, Long Backs, Batchlers, Wivetts, Short Twelves, Long Twelves, Jenny why Gettest thou, Rogue why Winkest thou. The shortest slate was about four inches, and the others increased in length by inches, 'sometimes less or more as the workman pleaseth.'

In South Yorkshire there has been a tendency to gradually increase the size of the stone slates, and the writer only knows of one roof, now very dilapidated, with small old slates as previously illustrated and described from Wales. The nature of the stone from which the Cotswold slates are made prevented the increase in size which took place in Yorkshire and elsewhere.

The material progress of the past century was unhappily accompanied by a progressive decline in the artistic appearance of the ordinary or vernacular buildings—a decline which is still proceeding, and from which the roofing materials have not escaped. The appearance of a modern roof of Welsh slates, rigid, flat, and uninteresting, will not bear comparison with one of the older South Yorkshire roofs of 'grey slates.' The colours—greys and browns in rich variety—to which the stone slates weather are in harmony, unlike the Welsh slates, with the stonework of the buildings, with the rough stone walls of the patterned fields, and

[1] *Old Cottages, Farmhouses, etc. in the Cotswold District*, p. 46.

with the landscape as a whole, especially in the grey days of winter. The older builders lived nearer than we to Nature, from which all Art ultimately springs, and so their buildings were better in accord with the country in which they were placed. Not only does a roof of stone slates agree with its surroundings better than one of Welsh slates, but it is also better when regarded as a separate æsthetic object, for the thickness and the slight roughness of the surface of the grey slates give a texture to the whole roof surface. The mechanical exactitude of the wooden framework of the modern roof produces a dead level surface in the roof covering, while the old carpenters sought for strength and permanence rather than a finished appearance. Art is a product of other ages and cultures than ours and when it became uneconomic, it also became exotic and ceased to be natural.

The Romans in England fastened their roof tiles and slates with one iron nail, and many have been found on different sites. ' As the distance of the nail hole from the point varies considerably, even in the slabs of the same site, it is almost certain,' writes Mr John Ward, 'that they were fastened to boarded roofs[1].' But it may be suggested that the difference in the position of the nail holes would be equally suitable with a foundation of strong wattle work, the use of which was widespread in England and Wales. As an example: Alexander Neckam, who wrote in the twelfth century, in describing a hall of the period, said that the roofing materials were placed on wattlework ('crates').

The Romans used iron for their slate nails, but that material was too costly for ordinary buildings in later times, and wooden pegs were usual. Mr Thos. Winder has described the fixing of the stone slates at Oughtibridge Hall, a South Yorkshire building[2]. He says that the pegs appear to have been made by riving oak into sticks of twice the length required, whittling each end, and chopping the sticks across. The laths were of riven oak, and so could not be used in very long lengths. At Oughtibridge Hall they were about 3 ft. long, resting on three spars. The laths were $1\frac{1}{2}$ in. to 2 in. wide by $\frac{1}{4}$ in. to $\frac{1}{8}$ in. thick, fastened to the spars by three hand-made nails, and where the peg holes fell directly over a spar an iron spike was driven through the peg hole and into

---

[1] *Romano-British Buildings and Earthworks*, p. 264.
[2] *Builders' Journal*, III, p. 40 (February 25, 1896).

the spar. In the earliest buildings these nails and spikes were the
only iron used, except perhaps, says Mr Winder, a nail or two in
the door latch.

In the North of England the small bones of the legs of sheep
were used as pegs for the hanging of stone slates in the Middle
Ages, and the tines of stags' antlers are said to have been used for
the purpose at the fourteenth-century gatehouse of Worksop Priory,
on the border of Sherwood Forest: but oaken pegs, as described
above, were those chiefly used for the hanging of stone slates, and
also for the artificial burnt tiles.

Thorold Rogers thought that the preparation of tile and slate
pins was a by-industry of the woodmen, in which even their wives
and children were engaged[1]. At the present time, in Derbyshire,
the writer finds that the preparation of thatching pegs is a by-
industry of the woodmen, from whom the farmers buy the pegs in
bundles at the price of sixpence the hundred: but the writer has
been told by an old slater that in South Yorkshire the slate pegs
were prepared, not by the woodman, but by the slater's apprentice
during the winter, when work was 'slack.' Similarly, in South-West
Surrey, according to Mr Ralph Nevill, the making of tile pins was
valued by the bricklayer as a useful occupation for winter; there
hazel or willow was used for the pins, but 'the delight of the tiler
was to get hold of an old elder stump, which was supposed to make
the most durable pins of all[2].'

Stone slates and tiles were formerly bedded on vegetable
materials, such as hay and straw, a survival, probably, of the roof
covering of the earliest buildings: this method had also the practical
advantage of covering the joints of the slates or tiles and helping
to keep out the weather. In many districts moss was used which
was not the ordinary moss which grows on walls, but long bog-
moss, in appearance somewhat like the thin sea weeds found on
the sea shore at high water mark. The gathering of this moss
from the boggy places on the moors in South Yorkshire was the
work of the women.

The mossing of roofs was a special trade, and in North Wales
the man who went round periodically was called, in Welsh, the
moss man, and the slates themselves were known as moss stones[3].

---

[1] *History of Agriculture and Prices*, I, p. 491 and IV, p. 444.

[2] *Old Cottage and Domestic Architecture in South-West Surrey*, p. 33.

[3] H. L. North, *Old Churches of Arllechwedd*, p. 100.

In Yorkshire and in Derbyshire the man was known as a mosser, and Mr Addy has a notice of a 'moser' of Totley who was killed by falling from the roof of Bubnell Hall, near Chatsworth House, on July 21st, 1708. He also quotes an entry in an old memorandum book: 'May the 15, 1762.—Thomas Rodger has don 18 days worke of mosing at Newfield Green,' which has never been more than a small hamlet[1]. The original moss decayed in the course of time, and the mosser obtained his living by poking up new moss under the slates, to keep out draughts and snow: in Derbyshire this was done with a square ended heavy trowel known as a mossing iron. Mossing is now obsolete in Derbyshire, and the stone slates are pointed with mortar, in the same manner as the Welsh slates.

In the year 1477-8 the churchwardens of St Michael's, Bath, bought 'pack moss' for five pence, and they paid a tiler a like amount *per diem* for 'moseyng et poyntynge' a stable. In the year 1463-4 they had bought tiles and hay, the former being laid on the latter[2]. The *Memorials of Ripon* supply a northern example: in the year 1493-4 one penny was paid for one bundle ('cercina') of moss for use with slate stones[3].

In the fifteenth century hay and straw were bought to be laid under the slating of St Peter's Church, Oxford, and in the year 1517-18 the chamberlain of Leicester bought 'sklates' and 'sklat pynnes' and 'lytter,' and paid two slaters and one server in connection therewith[4]. Here the slates seem to have been bedded on a poor kind of straw.

According to Mr Curtis Green, in some parts of Surrey the practice of bedding roofing tiles on clean straw is still followed, but not for better class work[5].

The small stone slates used for roofing in West Somerset were pegged *over* the rafters (as was usual everywhere), the pegs being driven firmly into the slates, and projecting only on the underside. In order to keep the slates in their places, and also to prevent the wind from disturbing those of small size, the row of pins along

[1] S. O. Addy, *A Glossary of Words used in the neighbourhood of Sheffield, s.v.* Mosing, pp. 151 and 322.

[2] *Proceedings of the Somersetshire Archæological and Natural History Society,* 1877-80.

[3] III, p. 164.

[4] *Records of the Borough of Leicester,* edited by Miss M. Bateson, III, p. 9.

[5] *Old Cottages and Farmhouses in Surrey,* p. 28.

each lath was buried in a rim of mortar, which set around them and kept them firm. This operation was called 'pin pointing' or 'pin plastering[1].'

In the *Dictionary of the Architectural Publication Society* it is stated that formerly in fixing tiles, a piece of oak, somewhat triangular in section, about $2\frac{1}{4}$ in. or $2\frac{1}{2}$ in. long, was inserted in a hole from the back of the lath, and so prepared, was dropped on to the moss on the lath, usually one piece to each tile[2]. Unfortunately the compilers of this dictionary did not realise the importance of naming the districts where the methods which they described were in use, and the dictionary says, vaguely, 'Hay, straw, and moss were formerly, and in some parts of the country are still, used when laying tiles and slates, to assist in keeping out damp and cold, and especially drifting snow[3].'

In the year 1526 such materials were laid under lead at Oxford: but lead is beyond the scope of this chapter, as it. like iron, was almost a precious metal before cheap carriage of materials was brought about by improved methods of transit in the nineteenth century. Therefore in the old slate roofs the slater formed the valleys, not with lead, but by working the slates or tiles round them and the methods varied in different districts[4]. In stone slated roofs hips were almost unknown.

In the year 1813, Archdeacon Plymley wrote of the stone slated roofs of Shropshire, that they would be much firmer when put on if the ridges were covered with crests of stone, instead of turf, or perhaps no crests at all[5]. Turf is still used to form the ridges of thatched roofs in Derbyshire (Figs. 56 and 57).

The 'stone crests,' which Archdeacon Plymley desired, were in use in England for stone slated roofs in Roman times, and in Mr John Ward's *Romano-British Buildings and Earthworks*, there is an illustration of an example of Bath stone from Llantwit Major, in Glamorganshire[6].

Mr E. Guy Dawber has described how the stone ridges, like

---

[1] *English Dialect Dictionary*, s.v. Pin.

[2] s.v. Tile Pin.　　　　　　　　　　[3] s.v. Hay.

[4] Those of the Cotswolds and of Northamptonshire have been described by Mr Edwin Gunn in the second *Building Construction* volume of the Architects' Library, that of North Wales in the *Old Cottages of Snowdonia*, and that of North-West Yorkshire by Mr H. E. Henderson in the *Builder*, 23rd June, 1911.

[5] *General View of the Agriculture of Shropshire*, p. 106.

[6] pp. 263-4.

miniature roofs in shape, are economically sawn out of the stone in the Cotswolds[1]. Stone ridges of similar type are in general use with the 'grey slates' of the Pennines but there they were worked by hand out of the solid. Such ridges are costly, and Professor Thorold Rogers found, from the accounts which he examined, that artificial tiles were used for the ridges of stone slated roofs in the Middle Ages[2]. The modern blue Staffordshire ridge tiles without flanges are evidently copies of the old stone ridges. The stone slated roofs of Yorkshire were rarely hipped, but sometimes they were and the angle was then covered with stones similar to the ridges.

The old method of forming the ridges of slated roofs in the English Lake district has been brought to the writer's notice by Mr W. G. Collingwood. The slates used (of course, 'Westmorlands') were nailed *at the bottom*: they projected above the ridge, and were laid alternately on each side of the roof. On every slate a notch was cut out on each side at the top, and each notch fitted into the similar notch of the adjoining slate. The notches thus locked the slates together, and from this was derived their local name of 'wrestlers[3].' They seem to have been used as early as the time of Queen Elizabeth. The 'wrestlers' cannot have made a very satisfactory ridge, but the builders had, as was usual, to make the best of the local materials in the absence of facilities for carriage, for in the Lakes there were no tiles, and no stone suitable for working into those ridges, described above, like miniature roofs, which were used in the country to the south.

In North Wales the usual ridging of the local slating was at first probably made of clay, and at a later date of lime and hair mortar, and neatly whitewashed; the method was probably borrowed from thatching, as was the laying of the slates on wattlework[4].

The twelfth century writer, Alexander Neckam, says that a hall might be roofed with straw, rushes, shingles, or tiles (' chaume,

---

[1] *Old Cottages, Farmhouses, etc. in the Cotswold District*, p. 48.   Mr Dawber has given illustrations of the sawn stone, Figs. 60 and 61.

[2] *History of Agriculture and Prices*, IV, p. 443.

[3] Mr Collingwood has kindly presented a specimen to the writer. The method seems more applicable to wooden shingles than to slates, and may have been derived from them. Shingles seem to have been formerly used for church roofs in the Diocese of Carlisle.

[4] H. Hughes and H. L. North, *Old Cottages of Snowdonia*, p. 61.

ros, cengles, teules'). King Henry III in the year 1260 gave
orders for the thatch to be taken off the outer chamber in the high
tower of Marlborough Castle, and replaced with shingles[1], which
are pieces of wood of the form of tiles and slates, and laid in
a similar manner. R. Holme in the year 1688, defined shingling
as a mode of roofing with cleft wood, about 6 or 8 in. broad
and 12 in. long, 'pinned at one end to hang on the Laths.
They are laid as Slates with Moss under them, which is termed
Mouseing[2].'

The manuscripts of the early Middle Ages contain illustrations
of roofs covered with shingles of the same shapes as the Roman
stone slates. In a MS. illustration of about the year 1120 the
shingles are being nailed by the workman below the lap, but this
may have been an error on the part of the draughtsman[3]. Shingles
were very much used, even on important buildings, in the earlier
Middle Ages: thus, Salisbury Cathedral was at first roofed with
shingles from the Bramshaw woods in the New Forest[4], and in
1281 twelve oak trees from Sherwood Forest were sent to the
Franciscan Friars of Lincoln for shingles. The masons copied
them for the stone ornaments of buttresses, etc. These are more
common on the Continent than in England, but examples occur
at Lincoln Cathedral, in work of the thirteenth century.

The use of the stone slates must be regarded as an attempt to
make a roof of shingles in a more permanent material than wood,
and, in addition, slates had in some places the advantage of a
greater cheapness, which has ever been one of the motive forces
in the development of our building construction. Much of the
solidity of our old buildings was very largely due to the builders'
lack of experience of the correct balance between strength and
durability. Robert Reyce, writing in the year 1618, of the mansions
that were then being erected in Suffolk, complained that 'in all
buildings this one thing is observed, spare of stuffe, scarcity of
timber (which is too general), and that workman that can doe his
worke with most beauty, least charge albeit nott so strong, he is

---

[1] T. Hudson Turner, *Domestic Architecture in England from the Conquest to the end
of the Thirteenth Century*, p. 251.

[2] *Academy of Armory and Blazon*, p. 266.

[3] The illustration is reproduced by T. Hudson Turner in his *Domestic Architecture in
England from the Conquest to the end of the Thirteenth Century*, on the plate opposite
p. 8.

[4] *Dictionary of the Architectural Publication Society*, *s.v.* Shingle.

most required[1].' Nevertheless, to us, the buildings of the early seventeenth century appear exceedingly solid and well built.

In the year 1314 it was found that certain of the royal manor houses and castles which were roofed with wooden shingles were greatly in need of repair, and that they might be roofed at a less cost with stone slates or earthen tiles[2]. Professor Thorold Rogers found that the use of shingles hardly lasted after the fourteenth century[3], and J. H. Parker also noticed a decline in this method of roofing in the same century, which was a period of great improvement in social conditions.

In England the use of shingles is now confined to such structures as church spires, and those only in the South-Eastern counties, but the English in America used them for roofing all kinds of buildings. They are still used there and made of various timbers, such as redwood, cypress and cedar, and are sawn about 16 in. long and in random widths from $2\frac{1}{2}$ in. to 14 in.

It is not easy in the mediæval building accounts to tell whether 'sclats' or slates are of wood, stone, or tiles. In the churchwardens' accounts of St Michael's, Bath, twelve pence is paid for 'sclattyng' a house floor, and at the present time we use the word 'slats' for short, flat pieces of wood. In many modern dialects the word 'slats' is applied to roofing slates, as at Mansfield, in Nottinghamshire, and in the Cotswolds. In both of these localities the term is applied to stone slates. It would appear that the early builders did not distinguish between wooden and stone slates, and similarly, the Irish word 'slinn,' which is now applied to slates, was formerly applied to shingles.

The use of stone slates and of artificial tiles in this country extends to the beginning of the historic period—that is, the time of the Roman occupation—and it is unsafe to assume that the use of either was copied from the other.

In the accounts of Finchale Priory[4], the Latin word 'tegula,' that is, a tile, is applied to stone slates from Esh, while in a fifteenth century agreement for the rebuilding of a house at 'Haldisworth Inge,' near Halifax, it is stipulated to be covered

[1] *Breviary of Suffolk*, p. 50.
[2] J. H. Parker, *Domestic Architecture in England from Edward I to Richard II*, p. 8.
[3] *History of Agriculture and Prices*, I, p 492.
[4] *Priory of Finchale*, published by the Surtees Society, p. ccccl.

'cum tegulis Anglice sclatestones[1]' and in the contemporary *Catholicon Anglicum* 'a telestane' that is, a tile stone, is translated 'tegula,' and 'to tele' is 'tegulare, tegulis operire,' meaning to tile, to work with tiles; and in the same century in the *Promptorium Parvulorum* 'ymbrex,' another Latin word for roof tile, is given as the equivalent of 'slatstone.'

The history of the manufacture of tiles for roofs in this country is obscure. The roof tiles used by the Romans in Britain were of Italian pattern and left no descendants in later times, and it is uncertain when the manufacture of the plain tiles used in the mediæval buildings of England was begun. The Statute of Edward IV in the year 1477, which regulated the sizes of bricks, also regulated the sizes of tiles. The size of each was to be at least $10\frac{1}{2}$ in. by $6\frac{1}{4}$ in. by $\frac{5}{8}$ in. This size was retained by Acts of George I and George II several centuries later[2]. Probably the old tiles were baked not far away from the place where they were to be used, but the local character of tiling has been destroyed in modern times by the cheapness produced by manufacture on a large scale and the modern facilities for carriage, which has permitted the manufacturers, in places favoured by Nature, to send their goods all over the country, and so squeeze out the small local manufacturers. In the suburbs of any large modern town it is now possible to see in one road tiles from different English manufactories; on one house they may be of a bright red colour, while on another and more recent house the tiles may have been quickly toned by the weather to a fictitious appearance of age.

An extended use of tiles accompanied the renaissance of brick-work in the so-called Queen Anne style, which was based on the domestic architecture of the counties about London. The renaissance of brickwork led also to the use of tiles in districts to which they are unsuited by climate, etc., as the borders of the West Yorkshire moorlands. Tiles were formerly a vernacular material, but even in South-West Surrey they were preceded by thatch and Horsham slates, as Mr Ralph Nevill has shown. Now they are used in the best domestic work, and considerable care is taken in the manufacture to produce an 'artistic' appearance. It is doubtful whether the old craftsmen did more than attempt to

[1] H. Ling Roth, *Yorkshire Coiners*, p. 155.
[2] *Dictionary of the Architectural Publication Society, s.v.* Plain Tile.

make a good weathering tile, and whether the artistic appearance
of the old work was not due, partly to an instinctive and
unreasoning desire to make the tile pleasing in appearance, but
also partly, and more probably, to the hampering of the workman
by crude methods of manufacture.

The waved tiles known as pantiles were a comparatively
recent introduction from the Continent.   According to the
Dictionary of the Architectural Publication Society, before the
time of Daniel De Foe they had been imported from the
Netherlands: he is said to have introduced their manufacture
at Tilbury in Essex, but his venture was not a financial success.
An Act of Parliament, of the year 1722 (12 George I), specified
that ' All pan-tiles made for sale in any part of England, shall,
when burnt, be not less than $13\frac{1}{2}$ in. long, and not less than $9\frac{1}{2}$ in.
wide, and not less than $\frac{1}{2}$ in. thick[1].' They continued to be imported until past the middle of the eighteenth century, but pantiles
of English manufacture succeeded them, and their use was
becoming general when William Marshall published his *Rural
Economy of Yorkshire* in the year 1796. He noticed on his
journey to Yorkshire that there was not one roof of pantiles
between London and Grantham, but 'north of Grantham they
are become the almost universal covering[2],' and another topographer, in the year 1800, noticed that pantiles were taking the
place of thatch in North Yorkshire[3]. In this age of revivals
pantiles have recently come into use again.

We have seen how the stone slates have been ousted by the
Welsh slates for ordinary work, and that it is probable that the
stone slates, in their turn, had been a substitute for the earlier roof
covering of wooden shingles. In the following chapter, thatching,
a still older kind of roofing is described.

---

[1] *Dictionary of the Architectural Publication Society*, *s.v.* Pan Tile.

[2] I, p. 100.

[3] John Tuke, *General View of the Agriculture of the North Riding of Yorkshire*, p. 35.

CHAPTER XIII

THATCHING

Objections to thatch—Straw and its preparation for thatching—Different methods of securing thatch on the roof—Rods and broaches—Their names in the dialects—Scraws—Rope thatch—The thatcher's tools and their variety—The ridge—'Anointing'—Thatch of heather, rushes, etc.— European thatch—Thatching in the documents—Development.

Thatching is now but rarely used, and so less has been written of it than of any other of the old building crafts. The Modern Bye-Laws forbid its use, the makers of text-books on building construction have left it severely alone, and yet it is, in many ways, the most interesting of all the building crafts. Thatching has only quite recently felt the influences of printing and improved communications, the two great destroyers of local ways of work and thought. The capitalist thatcher, working in any part of the country, as required, has been developed quite recently, and the writer is not aware of any widely-circulated paper on the craft before the Board of Agriculture and Fisheries published their little gratis leaflet, No. 236, in June 1910. The tendency of such a publication must be to destroy those local aspects of the craft which make it so interesting to students of old building construction, for thatching is still carried on as trades were in the generally deplored days of tradition, before the introduction of machinery, and books, and certificates in building construction, in the times when the workman made his own tools, instead of using standardised tools procured from a shop. This primitiveness of the trade forms an obstacle to the investigation of its development, as the thatcher was a humble worker, working on his own lines[1], and leaving few

[1] In one small village near Cambridge, in August 1915, the writer found two thatchers whose thatching methods were different ; thus one laid the straw over the ridge, the other against the ridge, and so on. The variety in thatching, as shown in this chapter, is unknown in carpentry or bricklaying, but possibly this Cambridgeshire village may be a meeting place of Midland and East Anglian methods. The geographical distribution of English thatching methods has not been worked out. (See also p. 156).

written records behind him.   In such investigation the old work
is more valuable than the old records, and we have seen that it
is still possible to examine the work of the carpenter of many
centuries ago, but here, in studying the history of thatching, another
difficulty is met with, for an examination of old thatch is impossible, as it is very unlikely that there is any complete thatch
existing to-day which is as much as a century old, even if no
allowance is made for the 'nothing-like-leather' spirit in which
thatchers speak of the durability of their form of roof covering.
It is, therefore, impossible to apply to the investigation of its
development, the actual examination that can be followed in the
study of old stonework and old woodwork, and it has seemed
best to take the oldest description of thatching known to the
writer and to use it as the foundation for a more complete account.

This oldest account in detail of English thatching is that of
Henry Best, a farmer, of Elmswell, in East Yorkshire, who in the
year 1641 wrote instructions, always rambling and occasionally
obscure, for the thatching of cottages and stacks.   Some of the
instructions have unfortunately been lost : those that remain have
been published, with others, by the Surtees Society[1].

It is necessary to remember that thatching is a 'decaying
industry' at the present time, and also that the craftsmanship in
a trade which is declining is not usually as good as it was in the
days of its prosperity.   The industry is decaying because thatch is
an unsuccessful competitor as a roofing material with slates and
tiles.   This competition was naturally first felt in districts where
the slates and tiles were easily obtained.   We have seen that tiles
superseded thatch in South-West Surrey, and in the year 1805,
Messrs J. Bailey and G. Culley found that thatch which 'used to
be the universal covering,' had nearly fallen into disuse for cottage
roofs in Northumberland[2].   A few years before, in the year 1796,
W. Pitt had stated the objections to thatch, which was still
in use in Staffordshire[3] on roofs which were very pointed and
deemed too weak to support tiles.   These objections were that it
was liable to be torn off by storms, formed a harbour for vermin,
robbed the land of manure, and in dry seasons rapidly caught

[1] Under the title of Best's *Farming Book*, or *Rural Economy in Yorkshire* in 1641.
The quotations are from pp. 137–148, unless otherwise stated.
[2] *General View of the Agriculture of the County of Northumberland*, p. 27.
[3] *General View of the Agriculture of the County of Stafford*, pp. 21-2.

fire. Almost the whole of a considerable village, Wheaten Aston, had recently been destroyed through a thatch catching fire and spreading throughout the village, for when the thatch is on fire lumps of blazing straw fly in all directions. Its readiness to catch fire had caused its use to be forbidden in cities many centuries earlier. The townspeople of Hull in the reign of Queen Elizabeth were forbidden to 'theake with straw, reade, or hay, or otherwise than with thacke tyle,' and earlier, in the year 1212, a London ordinance permitted tiles, shingles, lead, and boards, but forbade the use of thatch of various kinds, as reed, rushes, and straw ('arundine,...junco,...aliquo modo straminis neque stipula')[1].

One cause of the disuse of thatch was the high price of straw at the time of the expansion of English agriculture a century or so ago, and at the present time in Surrey, according to Mr Curtis Green, its disuse is chiefly due to machine reaping, which damages the fibres of the straw and renders it unfit for thatching[2].

Sentiment also came into the matter, for thatching was the oldest form of roof covering, and the rule that the oldest fashions are the least esteemed, in building construction as in dress, worked against it in favour of newer materials, such as Welsh slates, and in addition it was cheap and impermanent, true progress in building being in the direction of cost and permanence. Palsgrave in the year 1530 wrote 'I am but a poore man, sythe I can not tyle my house, I must be fayne to thacke it.' Mr Frank Garnett, writing of the houses of the Westmorland 'statesmen' three hundred years later, says that in such a home of slates they were only used when the owner could afford to buy them, and that thatching with rushes was much more common[3]. In the year 1880, Reydon Church, near Southwold, in East Suffolk, was tiled on the side of the roof next to the public road, but thatched on the other side where it could not be seen[4], as thatch was considered to be an inferior form of roof covering, and for many years English churches with thatched roofs seem to have interested the readers of *Notes and Queries*, the reason being, apparently, that thatch is thought to be too humble a form of roof covering to be used on churches[5].

[1] T. Hudson Turner, *Domestic Architecture in England from the Conquest to the end of the Thirteenth Century*, p. 282.

[2] *Old Cottages and Farmhouses in Surrey*, p. 24.

[3] *Westmorland Agriculture*, 1800–1900. [4] *Notes and Queries*, 8th S., VIII, p. 418.

[5] In the *Builder*, 2 June 1911, there is an illustration, from a late sixteenth century manuscript, of the temporary thatching of an unfinished large church or cathedral.

The most usual material for thatching is grass—either one of the cultivated species, such as wheat, barley, oats, and rye, or grass of natural growth, such as true reed, which is largely used in East Anglia and North Wales. The use of the uncultivated grasses is probably the earlier, and in the Lincolnshire sea-marsh the word 'thack' is used for grasses growing in the dikes, though they are now never used for thatching. Rye straw was esteemed the best of the cultivated grasses because it was the longest and strongest. It was grown by some farmers solely for thatching, as it withstood wind, sun, rain, frost, and snow, 'and hardly suffered decay.' Sometimes the top layer only was of rye.

The next best was wheat straw, although Henry Best considered that the one was as good as the other, and wrote : 'Yow neede make no reckoninge which of those two it bee, for there is noe difference, but onely that rye strawe is the more usuall, if it bee to bee had[1].' However, in another part of the instructions he said, 'After that an howse is latted, the first thatch that is layd on would bee of rye strawe, well wroten amongst and well watered.' For the straw to be 'well wroten amongst' it should have been allowed to lie in some place for a time where pigs could rummage amongst it: this prevented it from growing on the houses. Best considered that 'haver,' that is oat, straw was good for thatch, but the birds would 'not let it alone.' If used it was to be 'anointed' with mingled water and lime like putty, with a trowel as the thatcher came down the roof. Best said that 'barley strawe is good alsoe, if it bee without weedes and not over shorte.'

The straw used in thatching should not have been threshed, as that destroys the continuity of the fibres, and in the past there seem to have been two methods by which unthreshed straw was obtained. In the older method the ears only of the corn were reaped, generally with a sickle, and tall stubble was left standing which was afterwards mowed for thatching. In the Middle Ages such stubble was called 'haulm,' with the usual mediæval varieties of spelling. Tusser[2] says :

> The hawme is the straw of the wheat or the rie,
> Which once being reaped they mowe by and bie.

He also says that :

[1] *Farming Book*, p. 60.
[2] *Fiue Hundred Pointes of Good Husbandrie*, chap. LVII, stanzas 14 and 15.

In champion countrie a pleasure they take
To mowe up their hawme for to brewe and to bake,
And also it stands them insteade of their thak,
Which, being well inned, they cannot well lack.

In the year 1807, Thos. Rudge[1] found in Gloucestershire two
kinds of unthreshed straw for thatching, the first, which was considered the better, was obtained by cutting off the ears of the
corn after mowing: the ears were separately threshed, and the
unthreshed straw was cleared of weeds and short stuff by a kind
of iron toothed comb fastened in the wall: it was then tied in
sheaves.   Such straw was called 'helm' and was used in the Vale
of Berkeley and 'the Lower Vale.'   The other kind, which was of
an inferior quality, was used in the neighbourhood of Gloucester
and 'the Upper Vale': it was composed of stubble, which was the
ground stalk of the wheat, left by the reaper about a foot long and
mowed afterwards: it was apparently mixed with threshed straw.
Henry Best tells that in the year 1641 the stubble was so short
that he was glad to mix haver (oat) straw with it in the proportion of two parts of stubble to one part of haver.   The custom
of mowing the straw long and cutting off the ears afterwards is
much older than Rudge's time, for Fitzherbert, published in the
year 1534, wrote, ' In Sommersetshire about Zelcestre and Martok
they doo shere theyr wheate very lowe, and all the wheate strawe,
that they pourpose to make thacke of, they do not thresshe it, but
cutte of the eares and bynde it in sheues, and call it rede, and
therwith they thacke theyr houses[2].'

'Reed' is now a West of England word for unthreshed straw
prepared for threshing, and such 'reed' must not be confused with
the true reeds used in East Anglia.   Wm Stevenson found that
in Dorset the more ordinary farm houses, and the major part of
the buildings in the villages and even some of the towns, were
covered with reed thatch, which was composed of wheat straw
not bruised by the flail[3].   At an earlier time the word seems to
have had a wider geographical range, for the Durham Account
Roll for the year 1415–16 contains an item for the purchase of
a roofing of straw called 'rede' ('in tectura straminea vocat rede
empt').   Thatching terms have varying meanings, due, no doubt,

[1] *General View of the Agriculture of the County of Gloucester*, p. 46–7.
[2] *Book of Husbandry*, chap. XXVII.
[3] *General View of the Agriculture of the County of Dorset*, p. 85.

to absence of literary standardisation, and in Somerset and Gloucestershire at the present time, the cutting-off of the ears of wheat is called 'haulming.' Properly, haulm is stubble, and in Somerset in the year 1437 'helm bote' was the right of cutting haulm in a common field. In the *Promptorium Parvulorum* haulm or stubble was translated by 'stipula,' and according to Bishop Kennett, haulm was straw left in an 'esh' or 'grattan.' Ray, in the seventeenth century, correctly defined 'haulm or helm' as 'stubble gathered after the corn is inned.' In Essex, in the year 1811, wheat stubble was haulmed, that is it was mowed, immediately after harvest.

Straw is usually wetted before being used for thatching. Best said that if dry straw were used it made the hands sore and dulled the 'eiz knife'; he also said that the difference 'betwixt strawe that is layd on dry, and strawe that is layd on wette, is that the wette strawe coucheth better and beddes closer.' Best gave full instructions as to the watering of the straw in summer, and said, ' In summer time wee allwayes desse and water our strawe, but in winter time wee onely throwe it out, and the raines and wette that falls are sufficient without any wateringe, for (this yeare) wee threwe out all our barley strawe that was threshed betwixt that time we gotte all in and the 17th of November,' then the straw was well wetted by the swine and on '18th of November when the thatcher came, wee did noe more but sette one of the threshers with a forke to shake up all the best of it, and lye it on an heape togeather, and then sette one to drawe it out immediately, and it was very good thatch.'

In Norfolk, at the present time, it is said that if the straw is not wetted it will not 'lie': with this object a quantity of straw is placed in a heap and soaked with bucketfuls of water to soften it.

After the straw has been wetted it must be 'drawn,' that is, taken from the wet heap and laid in straight bundles. In Best's time this was done by one of the thatcher's two women helpers. The bundles were called 'bottles,' and the drawer was to be provided with dry oat straw wherewith to make bands to bind them. The word bottle is still used in the dialect of Cumberland. The *Memorials of Ripon*[1] show that the drawing of the straw was done by women in Yorkshire long before Best's time; in 1379-80

---

[1] III, p. 102.

they record a payment of 2*s*. 1*d*. for helping to draw straw and
for carrying water at the same time. Best's second woman helper
carried up the 'bottles' to the thatcher on the roof, and Sir H. R.
Haggard, writing of Norfolk, says that at the present time for
quick thatching, three hands are required, one to make ready
the bands and broaches, one to carry the bundles of straw up the
ladder, and one to lay and fasten them[1]. In the West of England
the preparation of the straw for thatch is called 'yelming,' from
the Anglo-Saxon word 'gilm,' a handful. We have seen that this
work was done with a comb in Gloucestershire a century ago: in
Bedfordshire it was done by drawing the straw forcibly under the
foot: in Devon and Somerset it was done sometimes by a power-
driven machine called a reed-maker, which combed out the short and
bruised stalks: in Dorset it was pulled through a frame, and this
was called 'reed-drawing.' Richard Jefferies wrote of a Wiltshire
thatcher that he 'is attended by a man to carry up the "yelms,"
and two or three women are busy "yelming," that is, separating the
straw, selecting the longest, and laying it level and parallel, damping
it with water, and preparing it for the yokes[2].' At Winslade, in
Hampshire, according to the Rev. A. Kelly, M.A., the wet
straw was drawn out in long handfuls called 'yelms' and this
was often done by the thatcher's wife[3], as was also the case in
the Middle Ages, according to Professor Thorold Rogers[4]. At
Winslade a dozen 'yelms' were placed in the thatching 'fork,'
made of two sticks fastened in the shape of a V, and when full of
straw tied with a piece of string on the top, and carried up by the
thatcher on his shoulders. At present, in North Leicestershire, the
thatcher himself prepares the straw, wetting it and picking out the
short pieces, and he carries it up to the roof in his apron. Henry
Best recorded that the server carried up four bottles at a time, but
if the straw was very wet and long she only took up three bottles:
they were tied together and carried up by an old halter or a piece
of old broken tether, and the thatcher, who began like a slater, at
the eaves, stuck 'doune' his needles, one at a little distance from
another, 'thereon to lay the bottles when the server bringeth them
up.' In Norfolk the bundle of straw is fastened and carried on the
roof by a rope, and it rests temporarily against broaches, which

---

[1] *A Farmer's Year*, p. 283.   [2] *Wild Life in a Southern County*, chap. VI.
[3] The Winslade thatch was described by Mr Kelly in the *Church Monthly*, 1900,
pp. 205–8 and pp. 241–2.   [4] *Six Centuries of Work and Wages*, p. 235.

have been stuck into the roof[1]. Bundles of straw thatch were
known in Norfolk as 'gavels,' which is translated in the *Promptorium* as 'geluma, manipulatum,' each with the meaning of
handful. The writer has found that they are known as 'yelms'
in Cambridgeshire. In Nottinghamshire they are known as 'bats':
and there the thatcher's helper carries the bundle up to him as it
is required. The bundles are called 'bunches' in the dialect of
East Lothian.

Not only does the thatcher begin to lay at the eaves, but
also like a slater, he works on the roof from left to right.
Henry Best's instructions for the laying-on of thatch are as
follows: 'It is a greate oversight in many thatchers, that when
they are to lye on a whole thatch, they make it thicke att the very
eize, then they doe make it thinne upwards: whereas, on the
contrary, they shoulde give it a good thicke coat up towards the
toppe, and lye on noe more att the eize, but just to turne raine,
and by this meanes will it shoote of wette better by farre, when
it is full and not as it weare sattled about the mid-side of the
howse.'

The thatcher was to be provided with 'a great many bandes
for sewinge of the thatch' that was first laid on: the bands were
usually made of the finest haver, that is, oat straw, but occasionally of 'staddle hay.' They were made by two persons, one
person to sit beside the straw and feed the band, another to
go backwards with the rake, 'to drawe forth and twyne the same,'
and after that they were to be twined together again, 'after the
manner of a two plette.' Best continued, 'wee usually make our
threshers make the bandes, providinge three or fower allwayes
before hand, according to the number of places wheare it is to
bee sewed': if the 'forkes' were 15 ft. or 16 ft. high, the thatch
was to be sewed three times in the height, and four times if the
'forkes' were 19 ft. or 20 ft. high. The first sewing was to be
as close to the wall-plate as they could get, the second and third
2 ft. below and 2 ft. above the side 'wivers' (purlins) respectively,
and the fourth 'aboute a yard or more belowe the rigge-tree,
goinge straight forward, and att a like distance, fasteninge it aboute
everie sparre as they goe, and allsoe sowinge once aboute a latte
ever betwixt sparre and sparre.'

---

[1] Sir H. Rider Haggard, *A Farmer's Year*, p. 283.

In the British Isles there are four principal methods of securing thatch on roofs.   In addition to the sewing as described above by Best, the thatch may be held by rods laid across it, or the thatching material may be thrust into or between turves, or lastly, it may be secured by ropes stretched over the surface, which is the most primitive: these methods may be used either alone, or in combination, and they will now be considered in detail.

At Winslade in Hampshire, according to the rector, the Rev. A. Kelly, M.A., in laying new thatch (which Best called ' a whole

Fig. 55.   Cottage at Hawarden, built on crucks.   The roof is thatched and the gables are partly hipped back.

thatch ') ' after the straw had been carried on to the roof, a bucket of water is thrown over it to press it down the more easily.   After the " yelms " are laid, a ledger, that is, a pointed stick, is thrust into the straw, the length of it being carried across three or four "yelms" and tied to the rafters at the opposite end.   Under this another ledger is pushed for a few inches, and after securing a like number of " yelms," tied down in the same way.   By this method the thatch is fixed firmly to the roof, and is prevented from being blown off by the wind.   As the ledgers are buried beneath the surface of the straw, they are concealed from sight : but, at the bottom of the thatch spars are usually left visible, sometimes straight' (as in Fig. 55, a cottage on crucks, with partially hipped gable at Hawarden), 'or crossed in diamond fashion, and by increasing the number of these a pretty effect may be obtained[1].' Two illustrations, from photographs, are shown of a Derbyshire cottage: on the one side the thatch is out of repair and the ledgers have become uncovered (Fig. 56). On the other the thatch is in good condition, and the ledgers—or rods as they are called in Derbyshire—cannot be seen (Fig. 57). At Winslade the repair of old

Fig. 56. A Derbyshire thatched cottage. The thatch is out of repair and the ledgers or rods, which hold down the thatch, are visible.

thatch by laying new thatch over it was called 'covering,' and did not require half as much straw as when the thatch was entirely renewed. It was done in the same way as laying a whole thatch: but as the rafters could not be reached, owing to the old thatch, five spars, a few inches apart, were driven into the old straw at every five 'yelms.'

At Winslade, and also in Derbyshire, the rods which held the thatch are tied down, but at Elmswell, the straw itself was tied down, as it was likewise, the writer has found, on cottages in

---

[1] 'Warm in Winter, Cool in Summer,' in the *Church Monthly, ut supra.*

Carnarvonshire. In North Yorkshire, according to Mr Harold
Henderson, the thatch of rye straw was tied with ling fibre to the
rounded saplings used as rafters and well pegged down[1].

This method of securing thatch by rods laid across it is the
second of those into which the writer has grouped English thatch
and is that most generally used in England. The rods, or 'ledgers,'
may be either tied or 'sewn' to the rafters, or they may be held
down by 'broaches,' which are pointed sticks, generally of hazel,

Fig. 57. Shows the other side of the roof of the Derbyshire thatched cottage, shown in the preceding
illustration. The roof has been repaired and the
ledgers or rods are covered with the straw.

split, twisted, and bent into the shape of a hairpin, and used as
staples over the horizontal rods or 'ledgers.' These bent pegs are
sometimes used to hold down the ropes in rope-thatch (the last of
the writer's four groups), and they have a singular variety of names.
'Broach' is their East Anglian name and Sir Rider Haggard has
described their preparation as follows: 'One of the men, seated
close by, draws from a bundle, which for the last few days has

---

[1] 'Minor Domestic Architecture of Yorkshire' in the *Builder*, 28 June, 1911.

been "tempering" in a pond, some of the broaches which were split a few months ago : these he sharpens at either end, and then, holding the broach in both hands, wrings it in the centre, to make it pliable[1].'

The word 'broach,' French in origin, is old in English use, for the fifteenth century *Promptorium Parvulorum* translates ' broche for a thacstare, firmaculum,' and the broaches have recently been defined as ' rods of sallow hazel or other tough and pliant wood, split, sharpened at each end, and bent in the middle like an old-fashioned hair-pin, used by thatchers to pierce and fix their work[2].'

The name 'spar' is used for them in the West and South of England with such variants as 'spear' in Devon and Cornwall, 'sparrow' in Surrey, the Isle of Wight, Dorset and Cornwall, and 'sparrod' in the Isle of Wight[3]. In Berkshire, across the Thames from Oxford, the writer has heard the name 'spree' used. In *Old West Surrey* the thatcher held his loose straw in a 'frail,' or 'strood' : his spars were split and dexterously twisted in the middle to fasten the straw rope that he cleverly twisted with a simple instrument called a wimble. 'The lowest course was finished with an ornamental double bordering of rods with a diagonal criss-cross pattern between, all neatly pegged and firmly held down by the spars[4].'

In some south-western counties, Gloucester, Dorset, Wilts, and Somerset, they are known as ' spikes,' with such variants as ' spiks ' and 'speaks[5],' and the twist at the angle of the bend is said to be done in order to give them a tendency to spring outwards when fixed, and so hold more securely. In English-speaking South Pembrokeshire, the so-called *Little England Beyond Wales*[6], the broaches or spars are called both 'tangs' and 'scollops.' This latter name is found in Irish as 'sgolb,' a thatching peg, and it seems to follow the so-called ' Celtic fringe,' under variants as ' scob,' ' scope,' ' scolp,' and perhaps ' scrobe,' in Scotland, Ireland and Northumberland. In the latter county the ' scoubs ' are

---

[1] *A Farmer's Year*, p. 283.
[2] Walter Rye, *Glossary of Words used in East Anglia*, s.v. Broach.
[3] *English Dialect Dictionary.*
[4] Miss G. Jekyll in her book, *Old West Surrey*, p. 209.
[5] *English Dialect Dictionary.*
In Mr Ed. Laws' book of that title ; Glossary.

described as short rods of hazel or other tough wood, sharpened at each end, and then bent into the form of staples. They were put over the 'temple-rods' of the thatch. Other names for them in the North country were 'withy-neck' and 'spelk': and in Cumberland, for the liberty to cut spelks in the lord's woods, a hen was paid, which was known as a 'spelk-hen.' The broaches or spars are called 'pricks' or 'prickers' in Northamptonshire, Warwickshire and Norfolk[1], and the word 'prick' occurs in the list of thatcher's terms in Randle Holme's *Academy of Armory*, published in the year 1688.

In Derbyshire, Worcestershire, Shropshire, and Herefordshire such bent twigs are called 'buckles.' A Derbyshire thatcher, known to the writer, prepares them at home and carries them to his work in a bundle, an evidence of the primitive nature of the thatcher's craft. As the horizontal rods are made of the same material as the broaches, the words 'springle' and 'sprinkling,' used for the rods in Shropshire, Herefordshire, Hertfordshire and Essex, have been transferred to the broaches in Derbyshire and Warwickshire. Mr S. O. Addy has recorded that at Crowland, in Lincolnshire, thatching pegs are called 'speets,' from the Old Norse word 'spyta,' a stick or wooden spit[2], and the writer has found that the word 'spit' is used about Cambridge for both rods and broaches.

In thatching, there is a curious variety in the names used for the same thing in the same district. Thus, in Yorkshire, the thatching broaches are known as 'brods,' 'prods,' 'pricks,' and 'stobs.' This must be, in some degree, due to absence of literary standardisation, and possibly also to a coalescence of use in what was originally distinct; and the widespread use of the staple shaped pegs (broaches, etc.) seems to imply that old English thatch was not laid on laths, but on some material into which the broaches could be driven, such as sods as in the North, or brush-wood or wattlework as in Wales, at the present time.

The horizontal rods, by which the thatch is held down, also have various names. In Northumberland the prefix 'temple' is applied to them, and, in passing, it may be remarked that the Roman 'templa' are supposed to have been similar to our

---

[1] These dialect names, unless the contrary is stated, are from the *English Dialect Dictionary*.

[2] *Notes and Queries*, 9th Series, VIII, p. 178.

purlin. In Sussex the rods are known as 'radlings,' and as 'ledgers' in the Isle of Wight, Hampshire (at Winslade), and Dorset. In West Somerset the ledgers are described as 'laid horizontally across the row of "reed," and then tightly bound with cord, or more commonly withies, to the rafters : the durability of the thatch greatly depends upon the ledger[1].'

The third method of securing the thatch is by means of sods, called 'scraws' in the North, which also provide a foundation for the thatch. In one variety a small quantity of straw is pulled out at one end of the 'yealm,' turned down, and wound round the top of the 'yealm,' forming what is known as a 'staple.' With a thatching iron, an instrument slightly forked at the apex, the twisted head of the staple of straw is pushed through the turf, and is prevented from coming out again by the 'head' of wound straw. A Scottish form of thatching on turves, known as 'stob thack,' has been described as follows : 'The rafters are laid far distant from each other, on the coupling, and these rafters are then covered with shrubs, generally broom, laid across the rafters at right angles ; over this is placed a complete covering of *divots* (turf), which is again covered with straw, bound up in large handfuls, one end of which is pushed between the divots ; this is placed so thick, as to form a covering from four to about eight inches deep, and, after being smoothly cut on the surface, forms a neat, warm, and durable roof[2].' This was the 'ha'' or dwelling house, as the farm buildings had the straw 'spread loose on the roofs, laid much thinner, and held on with straw ropes, crossing each other like the meshes of a net,' that is, rope thatch as described below.

In Northumberland, and also, as the writer has found, near Sheffield, there is a method which is intermediate between the second and third as above described. The straw is laid on turves, but is held down by rods, which in their turn are secured by 'scoubs' or broaches driven into the turves.

The fourth, and last, method of securing the thatch is by ropes, and in the North it is called 'rope-thatch.' The most primitive kind of thatch we have, in which the straw is heaped on the roof without any order, is secured by ropes ; and the writer believes it to be the oldest form in use in these islands. Such thatch on

---

[1] *English Dialect Dictionary.*

[2] By an anonymous writer 'Agrestis,' of How of Angus, *in Edinborough Magazine*, 1818, p. 127.

the houses or 'clachans' of Lewis, was fully described by Sir Arthur Mitchell in his *Past in the Present.*

In the Lewis 'clachans' there was also to be found the most primitive form of roof, in which the rafters are laid, like joists, horizontally from wall to wall. As this primitive roof was superseded by the pitched roof, and did not form a factor in the development of English building construction, it has not been thought necessary to mention it previously.

On the flat roof the straw was heaped so thickly that it had an outline more or less semicircular when seen from outside. The thatch was removed every year for the sake of the soot which it contained, as that was considered to be a valuable stimulating manure. The ropes laid over the straw were of heather, and they were held down by stones, as weights, swinging from their ends[1]. According to another account the ropes are of straw; there are no projecting eaves, and the tops of the walls are covered with sods; some rain finds its way through the loose straw, but the amount is said to be trifling[2].

Mr E. W. M. Corbett has kindly informed the writer that in the 'clachans' the roofs are composed of 'a mass of rushes and rough grass, laid on cross timbers and formed a curved surface, tied on with hay-bands.' This roof covering, he says, is removed periodically, and carried on the backs of the inhabitants, generally the women, to the crofts, where it is spread to serve as manure, for in the absence of chimneys, the smoke from the house fire finds its way through the roof, leaving the soot among the grass, etc. Perhaps the roof was formed as it was with that object. According to Sir Arthur Mitchell, in heavy rain there were drops of black water everywhere.

The straw on the roofs of the Manx cottages is not laid loosely on a flat roof, but on a sloping roof in the usual orderly layers. It is held down by straw ropes which are crossed at right angles at intervals of 12 in. or 18 in. 'The duration of this roofing is short, not lasting above two years': this is due to the decay of the straw ropes[3]. The ropes are called 'sugganes' and the same name is used in Scotland and Ireland. They are twisted by

---

[1] *Past in the Present*, p. 53.

[2] H. Whiteside Williams in *Reliquary*, VI (1900), pp. 75–6.

[3] B. Quayle, *General View of the Agriculture of the Isle of Man* (1794), p. 17. T. Quayle, *ibid.* (1819), p. 23.

Fig. 58. Thatch held down by crossed ropes at Kirk Maughold,
Isle of Man. In the foreground is St Maughold's Cross.

U-shaped willow rods from straw teased out evenly with the flail, and they are now secured by tying their ends round coarse slates projecting from the walls, but formerly they were secured by loose stones, hanging from the ends of the ropes, as in Lewis[1].

At the present time in South Carnarvonshire, the writer finds that only the stacks are covered with rope-thatch. The straw ropes which run across the stack are secured to wooden pegs in the sides of the stacks, and the ropes are made on a bent instrument, curved like the handle of a walking stick, and called a 'doll.' Under the straw ropes, at right angles to them and to the material (straw, reeds, etc.) of which the thatch is made, there are laid slighter ropes, which the writer learns from personal enquiry were formerly woven of strong grass from the damp parts of the fields. The thatch of houses was different, and it is said to have been tied with straw ropes to branches which formed the foundation[2]. The thatching was done by the farmers themselves, and was not a specialised trade.

Sir Rider Haggard has described the Norfolk usage as follows: 'When the broaches are prepared, the thatcher and his assistant go to a heap of coarse hay, which is handy. Here the assistant fixes a wisp of hay on to an instrument called a bond crank, that is, a bent iron with a hook at the end of it, and two hand pieces of elder wood, so arranged that by holding one in the left hand and turning the other with the right hand, the hook revolves and twists the hay into a long grass rope or bond, as it is deftly drawn from the heap by the thatcher. These bonds, when of sufficient length, he winds into a sort of rough spool, with a broach for the centre of it, much as a fishing line is wound on to a stick[3].' After thatching the rope, or 'bond,' is laid 'across the straw, fastening it in place by the doubled up broaches which in shape and purpose resemble rough hair pins.' It has already been shown that in old West

---

[1] A. Herbert, *The Isle of Man*, p. 243, and the personal observation of the writer. Projecting stones, however, were used at least as early as the fourteenth century, as Messrs P. M. C. Kermode and W. A. Herdman in their *Manks Antiquities* (2nd edition, p. 86) say that in a small church or 'keeill' at Ballahimmon, 'there were found two loose "Bwhid suggane," that is to say, stone pegs set in the walls, for fastening the thatch.'

[2] Messrs H. Hughes and H. L. North in their *Old Cottages of Snowdonia* (p. 45), say that the reed thatch (which they describe) of the district was fixed to 'wattling,' perhaps the branches of the writer's South Carnarvonshire informants who could not speak English well. The reeds were either tied with withes to the wattling, or pegged to it.

[3] *A Farmer's Year*, p. 283.

Surrey, the straw rope was also held by broaches, there called spars. At the present time, tar band and coconut rope are used in the neighbourhood of Sheffield on stacks, instead of ropes of hay and straw, and they are wound round a thatching peg, as in Norfolk : but instead of the staple-like arrangement of the broaches, the band, or rope, is twisted round the heads of the pegs which are driven into the thatch unbent.

In the time of Henry Best the thatch of the stacks seems to have been held down by straw ropes, or 'bands,' running over the ridge, and down each side, like the straw ropes on the Carnarvonshire stacks to-day. Best said that the bands should not be 'above halfe a yard asunder,' and the two bands next the ends of the stacks should have two pieces of wood tied in each of them in order to keep the ends of the stack from rising[1]. The rest of the bands were all to be made fast at the eaves ('eize'), but he did not say how that was done, nor did he mention the slighter ropes parallel to the ridge and eaves. His description of the twining of the ropes has already been given.

In Scotland the ropes were secured to a pin fixed in the top of the gable, named a 'craw prod,' from its user. According to Jamieson, in 1808, there was a belief that if it were so covered with drift on Candlemas Day that it could not be seen, the ensuing Spring would be a good one, if not the contrary would happen[2]. In some districts the ends of the thatch on buildings were held down by slender poles: they were called 'gable poles' in Northamptonshire and 'lugs' in Warwickshire and Shropshire.

The width of straw laid at one time without moving the ladder is generally called a 'course,' and it varies according to the reach of the thatcher's arm. Best says that a course is 'aboute halfe a yard or more than a foote,' and at Winslade, in Hampshire, it is said to be generally a yard. In the Midlands a course is called a 'stelch.'

After the thatch is secured on the roof, it is usual in England to make the surface flat and even by beating and combing. This is generally done with a thatcher's rake. At Winslade 'each "course" is combed down with the thatcher's rake as it is laid.' In Nottinghamshire the rake is a long piece of wood with teeth or pegs on one side ; with it the thatcher smooths the straw and he uses the back of it to beat the straw down. The instrument is

[1] *Farming Book*, p. 60.
[2] *Etymological Dictionary of the Scottish Language*, s.v. Prod.

in general use in England. In Wales the straw is merely beaten with a wooden beater called a 'dobren.' It has been described to the writer as 'a club of some weight of oak or ash, to beat and smooth the thatch with the flat lower surface.' In Germany, as in England, it is the aim of the thatcher to produce a flat and smooth surface, and this is attained by the use of the 'streichbrett,' or 'dachbrett' (Lit. 'stroke-board' or 'roof-board'), a board with indentations or channels ('kerben oder rinnen'). Its use is general from Holland and from South Sweden to the South Tyrol and to Lower Austria[1].

Henry Best, in his list of thatcher's tools, says that the rake with three or four teeth is for scratching 'of dirte and olde mortar.' The other tools in Best's list are two needles for sewing, an 'eiz knife' for cutting the eaves, a switching knife for cutting even and all alike, as the thatcher came down from the ridge, a slice 'wherewith he diggeth a way or passage and also striketh in the thatch,' and a trowel for laying on the mortar.

This was in East Yorkshire, in the year 1641, and nearly three centuries later, in September 1911, a Nottinghamshire thatcher showed his tools to the writer. They were (1) a piece of wood shaped something like a battledore with a square end, which was used to 'bat' the eaves. He said that he had 'no particular name' for it, and it was apparently of his own invention. A 'piece of wood used by thatchers to beat down thatch,' in counties as far apart as Devonshire and Lincolnshire and called a 'batting board,' is apparently an English version of the Welsh 'dobren.' (2) A 'knife,' which was made from a scythe blade sharpened to a point and fixed longitudinally to a handle : the thatcher said that he had bought it from an old 'thacker.' (3) A 'rake,' which the thatcher had made by driving nails in a row through a straight piece of wood : it had no handle. (4) Shears, which seemed to be ordinary garden shears. He also had with him 'thack-pegs' and tar band. Here was a worker in a building trade, with tools which were not standardised, and some of which were certainly of his own manufacture and possibly of his own invention. In this respect the thatcher is the most primitive of building workmen, and preserves for us the old custom which has elsewhere disappeared.

Richard Jefferies has described for us the apparatus of the

---

[1] K. Rhamm, *Altslawische Wohnung*, p. 211.

thatcher of Wiltshire, whom he considered to be 'the most important perhaps of the hamlet craftsmen[1].' He has 'leathern
pads for the knees, that he may be able to bear lengthened contact
against the wooden rungs of the ladder, his little club to drive in
the stakes, his shears to snip off the edges of the straw round
the eaves, his iron needle of gigantic size with which to pass the
tar cord through when thatching a shed, and his small sharp bill
hook to split out his thatching stakes. These are of willow, cut
from the pollard trees by the brook, and he sits on a stool in the
shed and splits them into three or four with the greatest dexterity,
giving his bill hook a twist this way and then that, and so guiding
the split in the direction required. Then holding it across his
knee, he cuts the point with a couple of blows, and casts the
finished stake aside upon the heap.' He has also a 'fork' or
'yoke,' which is the forked instrument for carrying the elms to the
roof: 'these yokes must be cut from boughs that have naturally
grown in the shape wanted, else they are not tough enough.' In
Wiltshire the fork was also called a 'yelm'—or 'elm-stick.' The
writer finds the fork called a 'bow' near Cambridge and it seems
to be used on the ground. The forked stick is called a 'groom'
in Hertfordshire, and the name 'gillet' is given in Oxfordshire
to the 'forked apparatus used by the thatcher for carrying the
elms up to the roof.' In the south-central counties of England,
from Northamptonshire, southwards, the 'groom' is a wooden
instrument used by thatchers to keep the bundles of straw on
the roof before they are fastened down: the long spike is thrust
into the stack or roof, and the straw is then placed between the
roof and the semicircular bow, which prevents it from falling
down. A thatcher's bow, fastened to the roof to hold the straw,
is called a 'reed-holder' in Gloucestershire and the 'West Country.'
A thatcher in Mr Thomas Hardy's novel, *Far from the Madding
Crowd*, has a 'ricking-rod, groom, or poignard, as it was in-
differently called: a long iron lance sharp at the extremity and
polished by handling.' In Oxfordshire, Dorset and Wiltshire, the
'rick-stick' is "a rod with a few teeth at one end, and an iron
point at the other, by which it can be stuck into the thatch when
not in actual use," which is to comb the thatch lightly over, after
the elms have been fastened with spicks or spars[2].

---

[1] *Wild Life in a Southern County*, chap. VI.

[2] The dialect names in the paragraph are from the *English Dialect Dictionary*.

There is therefore the same variety in the names of the tools as there is in the names of the broaches, and quite as remarkable is the use of the same word for different tools, as the 'gillet' and the 'groom' which have already been described. This is probably due to absence of standardisation by literary use, and also because the thatcher, as we have seen, often makes his own tools, adopting, and modifying, and even inventing, to suit his special needs.

An old Derbyshire thatcher described his tools to the writer in May 1911 ; he had shears, a needle, 'like a big darning needle,' two knives (of which one was used for cutting the eaves), a rake for combing the thatch, made from the head of an old hay rake, one or two other tools, and a 'buttress,' which was of wood, long, flat, and narrow, and with a cleft at one end. In repairing thatch, he used it to push the wisps of straw up into the holes. He had also an iron 'buttress' which had belonged to his father and from which the wooden tool had been copied. In North Yorkshire a thatcher's needle is called a 'sting' or 'teng.'

The tools used in the repair of thatch are as curiously varied as all other matters connected with the trade. Some of them are described in the following paragraph.

In East Anglia and East Lincolnshire, a 'legget' or 'letchit' is a tool formed from a piece of board about a foot square, scored like a curry comb, with a handle fixed at an acute angle, and used for pushing moss, etc., off the roof, and for patting the thatch down. The word occurs in *Apprentices' Tools in Norwich*, A.D. 1548–64. In Northumberland a 'spartle' is a wooden *spatula,* about the size and shape of the flat hand, used by thatchers for raising up old thatch, in order to insert fresh wisps in repairing a roof. In the North of Ireland a small double pointed flat stick with a T head is used for thrusting in the knots of straw in repairing a thatched roof, and is called 'spurtle.' In Scotland the 'spurtle' was a broad-mouthed stick. In Cheshire an instrument used for raising up portions of the old thatch on a house and inserting the ends of the new thatch in the holes so made is called a 'spittle': it is said to be 'almost like the blade of an oar, and has a cross handle, or cosp, by which it is held.' These various names are evidently derived from the Anglo-Saxon word 'spitel,' a blade or spade, and show how the tool has been developed in various districts. In Leicestershire and Northamptonshire the name 'stinge' is given to the repairing of thatch by pushing in new straw under the old, and the

flat wooden instrument used is called 'stinger' or 'stincher': it is
'about a foot long and six inches broad, with a slit at one end for
the hand.' In Northamptonshire the instrument is also called a
'battledore,' and on the eastern side of that county, and also in
Warwickshire, it is called a 'gillet[1].'

The old Derbyshire thatcher already mentioned pushes in the
broaches or 'buckles' with the palms of his hands, as is also done
at Winslade and near Cambridge; but in Wiltshire, Dorset, and
Devonshire the thatcher drives home the spars with a small mallet
called a 'beetle.' In Norfolk the broaches are driven in by means
of a thatching mallet, and the straw, as it is laid on, is tucked
in where necessary, and drawn smooth with an instrument called
a thatching comb[2].

The methods of forming the ridge in thatching present almost
as much variety as does the thatch on the slopes of the roof.
Henry Best said little about the ridges of thatched buildings
except that they were made of mortar, and added that one woman
filled the scuttles with mortar as the thatcher threw them down
empty, and the other woman carried them up when full, which is
information of small moment to have survived for nearly three
centuries. The ridge of 'mortar' is still in general use in the
North of England, as it is also in North Wales, and its use is
much older than Best's time, as the Ripon accounts for the year
1399–1400 contain an item 'For the wages of one man tempering
mortar' ('*lutum,*' literally mud) 'for ridging for the said house, 4*d.*[3]'

The ridges on Best's stacks or ricks were formed differently
from those on his thatched buildings, and he gave the following
description: The thatcher 'goeth up in height till hee come within
a foote of the toppe: but on the toppe of all hee layeth noe thatch,
but onely loose strawe, which hee calleth the rigginge, and then
doth hee twyne hey-bands and cast over the stacke to keepe the said
rigginge from blowinge away......Our usuall manner is for the fore-
man to rigge our stackes, and then is hee to have two to helpe him,'
viz., one to draw stubble and bottle, and the other to give him up
bottles and bands and 'make the bandes fast att the eize' (eaves),
'and his manner is, first to lay stubble crosse overthwart the ridge
of the stacke, that the raine may runne downe, and then upon that

---

[1] These dialect names are from the *English Dialect Dictionary.*
[2] Sir H. Rider Haggard, *A Farmer's Year,* p. 283.
[3] *Memorials of Ripon,* published by the Surtees Society, III, p. 130.

hee doth lye more stubble eaven on the toppe of the ridge, thereby supposinge that the bandes which goe crosse the stacke will have the more power to keepe it downe, and soe that which lyeth above to keepe that fast and firme which lyeth under it[1].'

In North Yorkshire, a thick layer of rye straw was placed on the ridge and pegged and laced down : the eaves projected about 18 in., and the thatch was cut clean, at right angles with the surface of the walling[2].

Often in the South of England, the rods or broaches are crossed

Fig. 59.   Typical straw thatch at Hinksey, near Oxford.

over the ridge, in order to hold it down, in the direction of the roof slope.   They are most visible when the thatch is old and thin (Fig. 59), and are perhaps allied to the so-called 'Dachreiter,' that is roof-riders, of the German thatch.   'Dachreiter' are short pieces of wood crossing at the ridge, and holding down the ridge thatch, and these in their turn, seem to be a survival of the method of holding down the thatch by long poles laid over it, crossing at the ridge, which is the peculiarly Russian method (see pp. 216–7).

---

[1] *Farming Book*, p. 60.
[2] Harold E. Henderson, 'Minor Domestic Architecture of Yorkshire' in the *Builder*, 23rd June, 1911.

The thatch at the ridge, the eaves, and the verges is that which is most liable to be damaged by wind, especially when it lies to the West and South, and it is usual to fasten rods or ledgers on the top of the thatch at these points, as shown in Fig. 57.

The rods are often arranged decoratively, and Mr G. Ll. Morris has given us particulars of the methods employed in various counties; the most decorative thatching, he says, is that to be found in Norfolk and Suffolk[1]. In Derbyshire, it was formerly usual to cover the verges with the old-fashioned 'mortar': the old thatcher before mentioned now uses the more modern material, Portland cement, as shown by the illustrations (Figs. 56 and 57).

Autumn is the usual season for thatching with straw, and so it was a thousand years ago, for, according to an Anglo-Saxon writer thatching was then done in August, September, and October[2]. In Henry Best's time thatching was generally begun at 'Allhallow-tide' and ceased at Martinmas. Best, with a characteristic York-shire blending of kindliness and shrewdness, said that the best time for thatching is three weeks or a month before 'yow beginne to cutte grasse,' for then the days are long and the barns are empty for sewing, and later in the season it 'will not gette a man heate in a frosty morninge, sittinge on the toppe of an house wheare the winde commeth to him on every side.' But in spite of Best's suggestion thatching is still usually done in Autumn, which shows that a long accumulation of experience has established the practice of a thousand years ago.

The thatch when laid was not always left in its natural state, and we have seen that Best 'anointed' his thatch of oat straw to preserve it from the attacks of birds. But thatch was limed for other reasons; so treated, it is more durable than common thatch, and it has been recommended for the preservation of shingles on a roof, that they 'should be covered with slacked lime put on thick once a year[3].' Clay is said to answer nearly as well for the purpose, and Dean Aldrich[4], of Oxford, the contemporary of Wren, said that the custom of forming roofs 'with clay beaten together with short straw...remains even now in cottages.' By a mediæval law, thatch was required to be whitewashed to make it

---

[1] *Old English Country Cottages*, p. 59, see also pp. 143, 151-4.

[2] Quoted in *Anglia*, IX, 261, 17.

[3] *Dictionary of the Architectural Publication Society*, s.v. Shingles.

[4] *Elements of Civil Architecture*, 3rd edition (1824), p. 86.

burn less readily ; in Wales, the custom has survived the obsolete
law and has even been unreasoningly transferred to the slate roofs.
As late as the year 1833, the church of Llanrug in Arvon was
whitewashed 'roof and all[1],' and on the coast of Cardigan Bay
in Mid-Wales, not only may whitewashed thatch still be seen,
but whitewashed slates also. Such a treatment of the slates is
apparently of no practical use but it has an æsthetic value, for a
whitewashed roof of Welsh slates, its original white toned to many
shades by the storms from sea and mountain, is more beautiful
than the same roof left in its natural colour. Lime was also
applied to the roof, as to the other parts of the building, in the
Middle Ages, for purposes of sanitation and cleanliness : but it is
necessary to remember that the difficulties of transport generally
prevented the use of lime in ordinary buildings in any quantity,
except within a day's journey or so from the place of its production.

As we have already seen, the thatch in use at the present time
is usually of grass, either wild, as reeds, or cultivated, as the straw
of various cereals : but up and down the country, generally on
minor buildings, there are scattered examples of thatch of other
materials, which are evidently survivals from a time when such
thatching was of more importance, and probably in regular use
over wide districts. Such materials include heather or ling, whin
or furze, fern or bracken, and rushes.

Heather or ling is the most common of these minor materials
for thatching, especially in the North, where it is more easily
obtained : it is very durable and does not rot like straw, but its
appearance is very sombre, as it soon turns black under the
influence of the weather. In Scotland the work of laying is
begun at the eaves and the heather is placed in layers, with the
roots of each twisted together, and so up to the top : and when
both sides are covered, the roots are secured to the ridging by
well wrought clay, which is generally covered with 'divots,' that
is turves[2]. At Edmundbyers in the county of Durham, two
or three feet of heather thatch are said to have formed the
roof covering of old cottages on crucks, at the end of the last
century[3]. In one district of South Yorkshire, a little heather thatch
is still used ; the work is not done by the thatchers but by the

[1] H. L. North, *Old Churches of Arllechwedd*, p. 83.
[2] *Dictionary of the Architectural Publication Society*, *s.v.* Thatch.
[3] Rev. W. Featherstonehaugh in *Archæologia Aeliana*, XXII, ii (1900).

besom-makers who lay it with the stalk upwards as in Scotland. This laying of heather by the besom-makers seems to carry heather thatch back to a time when the thatcher's trade had not become specialised. Mediæval records of heather thatch occur in the Jarrow Accounts for the year 1370, as in that year the monks paid ten shillings for one hundred 'travis' of heather for the covering-in of houses ('coopertura domorum'), and it may be noted, in the same year they bought a thousand slate stones for eleven shillings. A century later, in the year 1477, they paid one hundred and eight shillings for ('adquisicione') heather and turf ('lynge et dovet'). In the year 1467 the monks paid eight pounds three shillings and five pence for repairs to houses and walls with tiles (probably stone slates), laths, and brods (wooden pegs)[1].

An entry under the date August 30th, 1672, in T. Whittingham's Diary, says, 'Wheatley of Saiston is to thacke' (with heather or ling) 'Leonord's barn and compleate for 26s.: it is 18 yards long. He hath 12d. for earnest, and I to be at no loss either with watling, ridging, or serving for ling,' which shows that heather thatch was laid upon wattlework in the seventeenth century.

In Dumbartonshire, thatch is formed of the common heath (*Erica vulgaris*), laid on with the stems downwards[2]. Dr Samuel Johnson has given a general description of the houses of the Scottish nation in his time. The roof was covered with heath, held by ropes of twisted heath, of which the ends, reaching from the centre of the thatch to the top of the wall, were held firm by the weight of large stones; there was no chimney, but a hole in the roof only, which was not directly over the fire, so that the house was filled with smoke[3].

Allied to heather-thatch is that of whin or furze, which Tusser, in his *Fiue Hundred Pointes of Good Husbandrie*, says should be cut in June for 'thacke[4].'

At Winslade, in Hampshire, straw was most commonly used, but furze, and sedge from the river, were also used[5]. In Cambridgeshire

---

[1] *Inventories and Account Rolls of Jarrow and Monkwearmouth*, published by the Surtees Society, pp. 56, 118 and 120.

[2] *Dictionary of the Architectural Publication Society, s.v.* Thatch.

[3] *Journey to the Western Islands*, sub 'Lough Ness.' Dr Johnson also described the house-walls made of turves, and lined with stones or wattlework.

[4] Chap. LIII, st. 12.

[5] Rev. A. Kelly, 'Warm in Winter, Cool in Summer,' in *Church Monthly* 1900, pp. 205–8 and 241–2.

the sedge *Cladium mariscus* is the main, though uncultivated crop of the fen, as 'it makes a beautiful durable thatch': its sharp serrated edges keep away birds and rats[1].

The old Derbyshire thatcher before mentioned, told the writer that flax was formerly used for thatching in that county.

In the *Memorials of St Giles, Durham*[2], there is an item of seven pence paid in 1614 for one thrave of 'spartes,' which were dwarf rushes, and for the laying-on of them.

In Westmorland, according to Mr Frank Garnett (*Westmorland Agriculture*, 1800–1900), the houses of the old 'statesmen' used to be thatched with rushes a century ago, and in Ravenstonedale the people were not allowed to cut rushes for thatching before 'the first Tuesday after St Bartholomew's Day at twelve o'clock in the day.'

In Lancashire, in the year 1795, according to John Holt, it was held that fern made the best covering, being naturally dry, and not apt to ferment, like straw[3].

Randle Holme, in 1688, named 'the several ways of covering Houses or other Buildings' and defined thatching and sodding as follows: 'Thatching is to cover them with Straw, Ferne, Rushes or Gorst, which is bound and held together by Laths, Windings and Thatch Pricks, done by the Art of the Thatcher. Soding is the covering of little shourings and places of shade from Rain, with green Turfs or Grass Sods, or paring of the surface of Heathy Earth, which being laid on the Roof of a House keeps it dry[4].'

During one of the writer's rambles in search of older forms of building construction, a group of farm buildings was found at Greno Wood Head, near Sheffield, in which the house was covered with the stone or 'grey' slates so usual on the older buildings of the district: one of the outbuildings was thatched with straw laid on bracken, and another outbuilding was thatched with bracken covered with a layer of sods: branches of trees were laid over both of the thatched roofs. The tenants said that they had copied the straw on bracken from Derbyshire examples: but the laying on of sods and branches was their recent contrivance, and had been adopted, because of the effects of recent high winds.

---

[1] T. M. Hughes and M. C. Hughes, *Cambridgeshire*, p. 75.
[2] Published by the Surtees Society, p. 44.
[3] *General View of the Agriculture in the County of Lancaster*, p. 16.
[4] *Academy of Armory and Blazon*, p. 266.

Mr Thos. Winder states that the 'grey' slates at Penistone, South Yorkshire, are sometimes covered with sods[1].

About the year 1814 the word 'biggin' was applied in the mountainous districts of the West Riding of Yorkshire to a building, generally a hut, covered with mud or turf[2]. Such a use of the term implies that the ordinary building in the district had once been poor and mean, and covered with such materials. We have already seen that charcoal burners' huts and others are still covered with turves or sods, and that the latter have been used as a foundation for various kinds of thatch.

It will now be interesting to examine the continental methods of thatching, and compare them with those in use in the British Isles, in the manner already attempted with the other forms of construction. The late Herr K. Rhamm investigated the thatching methods of the Teutonic and Slav countries[3], and decided that a roofing of turves or sods is Scandinavian, that heather thatch is Danish—or more strictly Jutish—and that the thatch of Germany is of straw or reed.

Gudmundsson, who carefully studied the evidence in the Sagas for old Norse methods of building, considered that the sod roof was of early origin, and that the sods were laid upon boards. Nicolaysen, who examined the ordinary Norwegian buildings of his time, described the sod roof as the warmest and most durable form of roofing: the sods were usually placed in two layers, the upper layer having the roots upwards, and the lower having them downwards, on a layer of birch bark, with boards underneath all. Experience has shown that roofs of boards only are cold, and for this reason they are not used for dwelling-houses. In Swedish Norrland, the sods of Norway are replaced by a layer of birch bark, ten centimetres, or about four inches thick, which are held down by a layer of boards running up and down in the direction of the slope of the roof. In subsidiary buildings ('Nebengebauden') in northern Sweden and in Norway, the boards are replaced by poles ('Stangen') which are placed both over and underneath the bark. On the island of Oeland, Linnaeus found a double layer of sods used on the straw roof: in this case the straw was secured to the laths in bound bundles[4].

---

[1] *Builders' Journal*, III, p. 40 (February 25, 1896).
[2] List of Dialect words in *Archæologia*, XVII, p. 140.
[3] *Urzeitliche Bauernhofe*, pp. 583-6 and *Altslawische Wohnung*, pp. 205-225.
[4] K. Rhamm, *Urzeitliche Bauernhofe*, p. 585.

Rhamm found that there are three principal varieties of straw thatch in central Europe: either it is heaped without order on the roof and fastened by means of overhung poles or rods ('Stangen oder Gerten'), which is the Great Russian and Serbian mode of thatching: or, secondly, it is taken on to the roof in bundles, which are loosened and spread out in sheets, pressed closely to the roof by means of horizontal rods (the equivalents of our English rods or ledgers), which is the German method: and, thirdly and finally, there is the thatch of the West Slavs, in which the sheaves or bundles ('Schäuben') are fixed on the roof without being un-loosened[1].

Fig. 60. Russian wooden house, of 'block-house' construction, in the Government of Kazan. The thatch is held down by poles, crossing at the ridge.

According to Rhamm, in the 'old Danish lands,' the straw bundles of the ridge are held down by short rods, 'Dachreiter' or 'Wartrae,' crossed in pairs over the ridge. The rods in English thatch remind us of them (Fig. 59). In Jutland and Sleswick they are only used in the timbered eastern districts, otherwise the ridge is fastened by a layer of sods. On the island of Gothland the 'Dachreiter' extend the whole height of the roof, and help to hold down the thatch: they meet at the ridge in pairs and are held together by a cross piece ('Querholz'). Two illustrations are

[1] K. Rhamm, *Altslawische Wohnung*, pp. 205–6.

shown, of which one (Fig. 60), is a Great Russian house in the Government of Kazan, with the thatch held down by overhung poles. The other (Fig. 61) shows an Esthonian outbuilding, and is an example of the expansion beyond Germany of the German mode of thatching, with the ridge thatch secured by ' Dachreiter.' At the top of the hipped end is a form of owl-hole (' eine Art Ulenlok ')[1].

At Herning in Denmark the roofs are formed of alternate layers of heath and straw. Rhamm thought that heath thatch was a middle point (' Mittelglied ') in the transition from turf thatch to straw thatch, but considered that it was a question as to whether

Fig. 61. An outbuilding of ' block-house' construction in Esthonia.
The thatch at the ridge is held down by crossed rods.

this transition was to the German type of straw thatch, or to a more primitive kind with the loosely heaped straw held down by poles, as in Russia.

It may be possible to group Continental thatch ethnologically as Herr Rhamm did, for there is little change among the peasants from generation to generation, and the answer of certain Slav farmers to all suggestions of improvement is said to be ' our fathers have been so accustomed.' In England, where there has always probably been more freedom than elsewhere, and a consequent

[1] These two illustrations are reproduced from *Altslawische Wohnung* and *Urzeitliche Bauernhofe* by permission of the publishers, Messrs Vieweg and Sohn. The permission, of course, was obtained before the war.

change from age to age, thatching has undergone the same slow development as the other crafts from the earliest time up to the present, and a few years ago there died, in a Leicestershire workhouse, an old thatcher who was called the 'king of the thatchers,' from the improvements which he had introduced into his craft. Nearly all the continental varieties of thatch are still in use in the British Islands, a fact which gives to these various methods a cultural rather than an ethnological significance.

At a cottage at Treeton, South Yorkshire, now occupied by a worker in a coal-pit, but said to have been formerly the manor-house of the village, the writer found three different kinds of thatch. The oldest part of the cottage was built 'on crucks,' and it was covered with straw thatch laid on sods, and held with rods and broaches, the latter penetrating the sods. The newer part of the cottage was of the seventeenth or eighteenth century, and covered with straw thatch, laid on laths and sewn 'aboute everie sparre' as described by Henry Best. The whole of the thatch had been repaired quite recently with a covering of straw fixed in a third manner, by wires running across it, and twisted round unbent pegs driven into the older thatch beneath. These three kinds of thatch were evidently fixed by local thatchers, whose different methods in different centuries do not imply that they were of different races, but rather that they improved their methods with the passing years. If the straw is heaped on the roof without order both in Great Russia and in Hebridean Lewis, it is not because Gaels and Russians are the same people, but rather because they are in a parallel stage of culture, just as, all over the world, prehistoric man without distinction of race seems to have made tools of stone long before he made them of metal.

A similar development and change took place with the thatching methods during the Middle Ages. The churchwardens of the parish of St Michael, Bath, were the owners of house property, and the accounts of their expenditure are still in existence[1]. As they commence before the middle of the fourteenth century, they are as old as any which have come down to us. The houses were usually thatched, and the accounts show that the thatch was not always of the same materials, and that the varieties of material run roughly in periods. Before the year 1416 straw and 'spykys'

---

[1] Published in the *Proceedings of the Somersetshire Archæological and Natural History Society*, 1877–80.

were bought; and from 1416 to 1500 helm in addition. These materials were replaced by 'culms' and 'spines' from 1500 to 1540–2, the former having been bought by the dozen and the latter by the bundle, and at this time the thatcher is called 'culminar' and 'calumer,' in place of 'stipulator': from 1541–2 to 1564 helm and spikes occur again, but without straw. The 'helm' was evidently unthreshed straw, and the 'straw' may have been either threshed straw, or stubble. It is not easy to define the 'culms,' but their association with 'calumer' seems to imply some kind of reed. The spikes were used probably as rods and pegs : in the year 1414 the wardens paid for the making of them, but there is nothing to show whether the pegs were bent to form broaches or not. The accounts also tell how, in the year 1416, a clerk ('clericus') and his wife were paid for gathering straw, for binding it, and for serving the thatcher ('stipulator'). In the year 1420, the wardens bought an acre of 'thache' at Bathwick, and paid for bagging, bearing, carriage, and laying. In 1541–2 'a thachere and his manne' were paid fifteen pence for a 'daye and a halfe.'

Other mediæval accounts contain other materials and methods. In the year 1517 the Chamberlain of the Borough of Leicester bought 'stoble,' 'thacke' rope, litter, laths and nails, and paid for the 'thakeyng' of the same. He also bought a load of clay at the same time[1]. These materials were evidently for rope thatch, for which there is no evidence in the Bath accounts. The straw rope used for thatching in South Lancashire is still called 'thacking-rope,' and the clay might be used either to form the ridge, or for dressing the straw, and most probably for the latter. No doubt the litter was mixed with the clay.

As we have seen, the Ripon accounts[2] contain evidence for mediæval thatching in the North of England : an account for the year 1399 contains an item 'and for five score thraves of straw bought for roofing one tenement upon the Cornhill, and for the wages of one man thatching ('tegentis viz. thekand') the aforesaid house for five days' (at five pence the day). A woman was paid fifteen pence, at the same time, for serving the thatcher. Then come the items for mud for the ridge, already quoted, and a payment for drink for the workmen. The mediæval accounts were carefully kept, and nothing was purchased for this roof at

---

[1] *Records of the Borough of Leicester*, III, p. 9.
[2] *Memorials of Ripon*, published by the Surtees Society, III, p. 130.

Ripon, but straw and clay; no 'thacke' rope as at Leicester; no 'spykys' as at Bath. The thatch was therefore probably fastened with home-made ropes of straw, like the thatch of Henry Best's stacks in the same county. Another and earlier Ripon account contains an item for 'cc stralanes,' that is, straw ropes. The same account also has a payment for 'vi pragges ferri,' that is six iron prags: more than five centuries have passed and the old Derbyshire thatcher before mentioned still uses irons which he calls 'prags' to secure the thatch to the spars. The fourteenth century Ripon accounts also show payments made for drawing and serving.

Rope thatch is probably our oldest form, and the wired thatch of the Treeton cottage is its latest descendant. Ropes may have been used to secure other roofing materials besides thatch. In *Nial's Saga* some ropes lying on the ground are mentioned, which were often used to strengthen the roof. 'Then Mord said "Let us take the ropes and throw one end over the end of the carrying beams, but let us fasten the other end to these rocks, and twist them tight with levers, and so pull the roof off the hall."' In this way the whole roof was pulled off[1].

The *Promptorium Parvulorum* shows that the method of holding thatch down by pegs like staples is as old as the Middle Ages. The variety of the names of such pegs, although in part due to absence of literary standardisation, shows that their use was not general in early times, as does also the absence of difference in name between the straight or bent pegs and the horizontal rods or ledgers.

The oldest roofs were probably of rough brushwood, and we have seen that, in building, brushwood was developed into wattle-work, which was used in Germany for roofing. In England, the use of wattlework was continued as a foundation for thatch. Bede tells of the burning of a roof of wattles and thatch in the North of England in the year 642, which is the earliest literary evidence of English roofing materials known to the writer[2]. Wattling is still used in North Wales as a foundation for straw thatch, and the writer has seen brushwood similarly used in North Derbyshire: it was considered to be fit only for inferior work, and was said to be 'cuttings from the hedge bottoms.'

[1] *Nial's Saga* 1, 244 quoted in *Notes and Queries*, 4th Series, VII, p. 244.
[2] *Ecclesiastical History of the English People*, Bk III, chap. x.

From the foregoing survey it is possible to classify our usual English straw thatch as of Germanic type, but modified by the influences of insular isolation, and the roofing methods of Celts and Scandinavians.

About the thatcher himself very little information is available. The thatchers do not appear to have formed any gilds or companies like the masons and the carpenters, and the reason may be that thatching was never thoroughly specialised. The *Nova Legenda Anglie*, under the year 1374, mention 'one Alan, surnamed "Wastelare," of Wainfleet, a coverer ('cooperator') of houses with different kinds of material.' Mr Laurence A. Turner has written of a craftsman at Barsham, near Beccles, in East Anglia, who called himself 'thatcher and dauber,' as his father and grandfather had done before him[1]. 'Unfortunately for him,' says Mr Turner, 'the county council's by-laws do not allow him to practise his craft as dauber, so that he has to fall back entirely upon thatching for his means of livelihood,' but here again the building bye-laws restrict his work. At the church of Barsham, a thick coating of clay was found underneath the reed thatch. In these two instances, the one from the fourteenth century and the other from the nineteenth, the thatcher was also a wall-maker, and his relationship is shown with the wattle-worker; we are carried back to a time when the crafts as well as the parts of the building itself had not become specialised. A thatcher and his assistant are mentioned in the thirteenth century, but it does not therefore follow that they did nothing but thatch roofs of buildings. 'Thacker' and 'thackster' were old forms of thatcher: both mean 'roofer.'

The thatcher's appetite was proverbially large and Best said that thatchers had (in most places) sixpence a day and their food in summer, and in the shortest days of winter fourpence per day and their food: but he never gave above fourpence per day because the diet was different. He gave them breakfast at eight o'clock, dinner at noon, and supper at seven, and at each meal 'fowre services.' If they provided their own meals they were paid tenpence per day. As already mentioned, Best provided, and also paid, two women to help the thatcher.

Best said that 'Many will (after a geastinge manner) call the thatcher "hang strawe," and say to him—

[1] 'Decorative Plasterwork' in *The Arts Connected with Building*, p. 172.

Theaker, theaker, theake a spanne,
Come of your ladder and hang your man.'

And from the roof the man answered—

When my maister hayth thatched all his strawe
Hee will then come downe and hange him that sayeth so.

The thatcher was called 'hang straw' in Worcestershire as late as the middle of the nineteenth century.

The information available as to the methods of roofing, with materials other than slates, in this country is so meagre that its collection and publication are desirable before it is too late. The craft of the thatcher is peculiarly one in which the old methods of work have lasted with but little change to the present time: but it is unlikely that they will last much longer, and they should therefore be recorded before they vanish with the inevitable revolutions of civilisation[1].

---

[1] Thatched roofs, in addition to those named in the text, are shown in Figs. 7, 28, 49, 50 and 51.

The East Anglian method of thatching with reeds has been described by Mr Thos. Winder in *Farm Buildings*: that of North Wales by Messrs H. Hughes and H. L. North in *The Old Cottages of Snowdonia*.

CHAPTER XIV

DOORS

Furze and materials other than wood—Loose doors—Single board doors—
Battens or ledges and their descent—The panel door—The heck door—
The folding door—The door bar—Bolts and snecks—The harr and the
hinge—Bands and hooks.

Openings are necessary in a building for various purposes, as
for the occupiers to go in and out, for the admission of light and
air, and for the discharge of smoke. Of these openings the first
is the most important; the method of closing it, usually by a door,
is described in this chapter.

One of the most primitive forms of door—if such it can be
called—in historic times, was the bundle of furze which was used
in the cottages of the Isle of Man 'from the scarcity of wood in
the Island[1].'

The earliest doors were loose, and there was no fastening at all
to the door of the South Yorkshire charcoal burners' hut described
in the second chapter. When the structure was unoccupied, the
door was leaned against a post in front of the doorway outside, and
the foot of the door rested against two wooden pegs driven into
the ground, as shown in Fig. 62. When the men were inside, and
wished to close the door, they simply pulled it over from the
position shown, against the opening, for there was neither hinge
nor handle. The expression 'put t' duur i' t' hoile' (put the door
in the hole), used in Sheffield with the meaning of shut the door, is
a survival from the time when doors were loose. A post before
a doorway, for the door to be leaned against, was possibly used
in more permanent buildings, as in a late mediæval Latin and
English dictionary there occurs 'a stoup before a door.' In the
Manx 'keeills' or small early churches, 'two instances are

---

[1] A. W. Moore, *Folk-Lore of the Isle of Man*, p. iv.

known, Ballacarnane, Michael, and Ballakilpheric, Rushen, where large stone pillars were set in front of the doors[1].'

It has already been shown that brushwood is easily developed into wattlework, and this was much used for doors. Until recently the doors of the South Yorkshire charcoal burners' huts were made of wattlework, and those of the bark peelers' huts of High Furness are still so made; and wattlework doors to more permanent buildings were in use in Cumberland in the eighteenth century. A

Fig. 62. Door of the South Yorkshire charcoal burners' hut shown in Fig. 2. It has no fastening, and when 'open' is simply leaned against a post fixed in the ground.

century ago the gates in the Cotswolds were described as 'little more than strong hurdles, made of split ash, or willow, with little workmanship or skill[2].' In Scandinavia, in the Middle Ages, the doors of the houses of the 'coloni' (villeins, or cottars), were 'hurdles of twigs and sticks (vændredor)[3].'

It is unlikely that doors of wooden boards or battens were

[1] P. M. C. Kermode and W. A. Herdman, *Manks Antiquities*, 2nd edition, p. 91.
[2] Thos. Rudge, *General View of the Agriculture of the County of Gloucester*, p. 100.
[3] Information kindly supplied to the writer by Mr Bernhard Olsen.

usual for ordinary buildings in England in the Middle Ages—at any rate, not in the less civilised North. In the stirring and well-known ballad concerning Adam Bell, Clym of the Clough, and William of Cloudesley, when William was fighting against heavy odds at Carlisle, to avoid arrest:

> There myght no man stand hys stroke,
>   So fersly on them he ran:
> Then they threw wyndowes and dores on him,
>   And so toke that good yeman.

The thrown 'wyndowes and dores' were evidently loose, and they must also have been of very light weight, for William was not harmed in the least: he and his two friends fought their way out of Carlisle, escaped, and then, having reached Inglis Wood,

> They set them doune and made good chere,
> And eate and drynke full well.

The miller of the *Canterbury Tales* broke doors by the simple method of butting them with his head.

In more recent times Sir Arthur Mitchell has described the doors of the 'black houses' of Lewis; they were commonly made of undressed wood, but he had seen large straw mats used as doors, and he had also seen doors made of a cow's skin stretched on a rough wooden frame[1]. In the year 1774, according to Thomas Pennant, the doors of 'skeelings' in the Hebrides, were made of birch twigs.

At a mud house at Great Hatfield, Mappleton, East Yorkshire[2], the original window, and also the door, were of harden, which is a kind of coarse sack cloth. 'The door could be lifted up like a curtain, and there was no inner door. Robbers were not feared.'

These old examples, from ordinary buildings, of doors of furze and birch twigs, of skins and straw mats, show that, for the doors, there was used the same variety of strange materials, far removed from modern practice, as we have already noticed in the other parts of the building, such as the walls and the floors, and it is evident that their use was continued until comparatively recent times.

In primitive structures the security given by a small door opening was preferred to the convenience of a large one, and

[1] *Past in the Present*, p. 54.
[2] S. O. Addy, *Evolution of the English House*, p. 41.

this is still the case in the buildings of savages, as in the houses at Malekula, New Hebrides, where the doorway is 'a small aperture ......only high enough to be crawled in at[1].' The small doorway survives in structures which are still made according to primitive models : thus the height of the doorway of the circular hut shown in Fig. 62 was three feet, and that of the High Furness hut, also described in an early chapter, was only ten inches higher. The doorway of such an important building as Greensted Church was only 4 ft. 5 in. high, and the doorways of the 'black houses' of Lewis were described by Sir Arthur Mitchell as 'very low, sometimes barely five feet high[2],' and Wm Horman, the author of an English-Latin phrase book, published in 1519, wrote, 'I hit my head against the soyle, or transumpt,' that is against the lintel. Mr Thos. Winder has noticed an old cottage at Ecclesfield Common, near Sheffield, where the only communication to the inner of two bedrooms was by means of an opening 2 ft. 6 in. high and 1 ft. 6 in. wide[3].

The old kinds of wooden doors are more interesting than doors of curious materials and the small size of door openings in early times, as wood is in universal use for ordinary doors at the present day.

Two doors were found in a timber structure in the Drumkelin Bog Crannog in the year 1833, which, at the time of their discovery, were believed to be of the 'Stone Age[4].' They were cut out of solid logs of oak, 4 in. thick and 2 ft. 7 in. broad by 4 ft. 6 in. long, with a piece of the solid wood projecting angularly at each end, a method of 'hanging' doors described on a later page. Such doors, made of a single plank, are the most simple, and therefore the most obvious, form of wooden doors; and examples of the Norman period existed in English churches until the 'restorations' of the nineteenth century. Doors formed of a board were also used on the Continent ; as an example, the two doors of Fortun Church, in Norway, are each of one pine plank 'about three feet in width[5],' which is wider than any plank of fir which can be obtained now.

The church in the Middle Ages was, as a rule, the finest

---

[1] Boyle T. Somerville, *Journal of the Anthropological Institute*, XXIII (1894), p. 373.
[2] *Past in the Present*, p. 54.
[3] *Builders' Journal*, III, p. 42 (February 25, 1896).
[4] *Archæologia*, XXVI, p. 367.   [5] *Notes and Queries*, 9th Series, X, p. 416.

building in the parish, and if its doors were of wood in Norman times, then the doors of ordinary buildings must have been of other and inferior materials. Such waifs and strays of building construction as the charcoal burners' huts and the 'black houses' of Lewis give us some clues as to what these materials were.

In later times, the wooden doors were formed of narrower upright boards, and it became necessary to fasten the boards to each other. This was usually done by the use of pieces of wood running in a horizontal direction across and at right angles to the upright boards. These are known as 'battens' or 'ledges,' and a door so made is named a 'batten' or 'ledged' door. It is the oldest form of wooden door constructed of more than one board, and it has been used from the earliest times up to the present. It is generally considered to be peculiarly mediæval, but Sir Christopher Wren used it for doors of small importance.

The batten door is a typical example of the descent in the social scale which is customary in methods of building, for in the Middle Ages it was used in the finest buildings, and is now described in the text books as the most inferior kind of door, suitable only for outhouses and similar buildings. The origin of the battens or ledges is obscure, but they were probably derived from a primitive method of securing a door of a single plank in its place. Palsgrave calls a ledge of a door 'a barre,' and we shall see that a bar was usually loose. Continental evidence also points in the same direction. In the year 1414 the churchwardens of St Michael's, Bath, paid fourteen pence for boards, three pence for nails, and one penny only for ledges, that is battens, for doors for three of the houses which were owned by the parish[1]. These doors of wood no doubt replaced doors of an inferior kind.

For the churchwardens of the same parish, in the year 1477–8, Galfrid the Carpenter pared, hewed, and planed four boards and five ledges for a new door for the tenement of Richard Lacey, in half a day, for which he was paid three pence. Half a hundred board nails were used in the same door. In this fifteenth century door at Bath we have evidence of a variety of door formed in two thicknesses by vertical and by horizontal boards. It may be regarded as a stronger development of the batten door, with the battens contiguous, and the principle of fixing two layers

---

[1] Accounts published in the *Proceedings of the Somersetshire Archæological and Natural History Society*, 1877–80.

of hardwoods, with the grain at right angles in order to avoid splitting, is well known to cabinet-makers. Mr Curtis Green has recorded the use of this form of door in Surrey[1], and the writer has noticed it in a seventeenth century house in Derbyshire, but in the country about Sheffield it lasted on into the eighteenth century. Two illustrations of these late doors are shown, of which one (Fig. 63) is from a photograph of the door of Midhope Church in the Little Don valley, dated on the stone head 1705, and the other

Fig. 63.   Porch and door of Midhope Church, South Yorkshire, A.D. 1705.   The panel in the gablet is of lead and the walls and 'slates' are of stone.

(Fig. 64) from a drawing to scale by the writer of a door at a farmhouse in the same neighbourhood, dated 1711. The mediæval appearance of these doors is apparent, and is evidence of the remoteness of the district from the stir of affairs. These doors also show the influence of the more important buildings on those of less consequence : the door of the church is the older, and it was copied six years later, with slight alterations, for the door of a farmhouse. When such doors were used as outer doors, the

[1] *Old Cottages and Farmhouses in Surrey*, pp. 40-1.

vertical boards were placed on the weather side, so that the rain did not rest easily in the joints.

The edges of the boards were simply 'butted' against each other in early batten doors, as at Langsett, in the Little Don valley, illustrated in Fig. 67. Then, later, the joints were covered

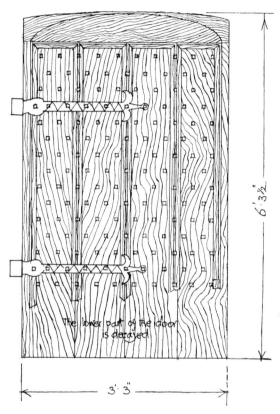

Fig. 64. Door of a farmhouse at Midhope, South Yorkshire. The door is made of double oak boards, fixed upright externally and horizontal internally. The bolts are of iron, copied from wooden pegs, as shown in Fig. 67.

with a fillet, which might be either chamfered or moulded; the former style was used at a house at Sharrow, Sheffield, of the middle of the seventeenth century, which was destroyed a few years ago; the latter in the doors at Midhope, shown in Figs. 63 and 64, and in the Surrey doors described by Mr Curtis Green, in his book on the old cottages and farmhouses of the county. At

Midhope the moulded strips cover the joints of the outer boarding as a protection from the weather, while in the Surrey doors they seem to cover the joints of the inner boarding.

The boards of the door at Langsett (Fig. 67) are fixed to the battens with wooden pegs, of which the heads project and are chamfered. Just as the winding stair of the carpenter was copied in stone by the mason, so such wooden pegs were copied by the smiths of the later Middle Ages, even to the chamfers, which were appropriate to wood but inappropriate to wrought iron; they are further evidence that the old craftsmen were not as much at home in the correct use of materials as it is the vogue to believe.

Batten doors have a tendency to drop downwards, or 'sag,' at the outer or fastening edge, and a piece of wood called a 'brace' was fixed from batten to batten, slanting diagonally upwards from the 'heel' of the inner or hanging edge of the door, to the upper batten. In a Danish barn door from North Zealand, illustrated in the *Billedbog fra Frilandsmuseet*, the hanging stile is the trunk of a tree, and a branch slanting up from the trunk forms the brace. Gates formed in this way were common in Lancashire a century ago[1], as they were at the same time in Dorset, where 'the head pieces by which the gates are hung, are sometimes forked, in which case one of the forks serves in some measure as a brace, to keep the gate from sinking[2],' but the writer does not know of any English doors of this kind, although they may survive in some remote places. As the early carpenters avoided joints, it is probable that the earlier form of braced door was like the Danish example. It is as unsatisfactory to study the development of our vernacular building construction without reference to the Continent, as it is to regard our architecture, in its history, as something apart from the contemporary work abroad.

The vertical board on the hanging side of the batten door was made thicker in order to strengthen it; in such old examples as remain in the Sheffield district, it is of the same thickness as the other boards and the battens together and the battens are tenoned into it. Such an arrangement may be regarded as the first step in the direction of the panel door now universally used, in which such

---

[1] John Tuke, *General View of the Agriculture of the North Riding of Yorkshire*, p. 100.

[2] William Stephenson, *General View of the Agriculture of the County of Dorset*, p. 174.

a thicker 'hanging' board is called a 'stile.' The further stages in the journey to the development of the complete panel door may be traced : first came the thickening of the other outer board or 'stile,' in order to carry the door 'fastenings,' and then came the placing of the battens at the upper and lower ends of the boards, and the running of the fillets horizontally on a board or 'rail' to form a decorative band, as at Midhope (Figs. 63 and 64). The last stages in the journey were the framing of the boards into the battens, which then became true 'rails,' followed by the recessing of the boards below the faces of the stiles and rails.

Doors are now generally panelled, except for very inferior work or in positions where it is desired to depart from the recognised and the conventional. Such doors are the cheapest that are made, and so their use has become universal in such ordinary buildings as are described in this book. We have seen that the timber is now converted into the required scantlings in the Baltic countries, but the exporters have proceeded further, and they now make joinery and send it over to England almost ready for fixing. The application of machinery to joiners' work reaches its greatest development in the manufacture of doors.

Those doors are still made by hand, of which the number required is not sufficiently large for the profitable use of machinery. Such are the doors divided either vertically or horizontally into two parts, which have been in use in England throughout the historic period.

'Heck-door' is the old and widespread name for a door divided horizontally in this manner, and in which each part is separately hung, and capable of being opened and shut independently of the other. Such a door had many advantages, as, for instance in houses without windows, the upper part, or upper 'heck,' could be opened for light : the door of the Southern charcoal burner's hut, mentioned on page 25, is a heck-door, and the only means of lighting the interior. Again, when a visitor whose intentions were doubtful came to a house with a heck-door, the lower part only might be opened, and if he then tried to force an entrance the difficulty of his position put him at the mercy of the householders : and the opening of the lower half only permitted of the going in-and-out of the smaller domestic animals, such as pigs and hens, which certainly lived in the English house up to the sixteenth century, and probably later.

Heck-doors were in widespread and early use on the Continent. In an old house at Ginheim-bei-Frankfort, shown in *Beilage zum Correspondenz-Blatte* for September, 1860, the door of the dwelling house and the door of the cow-house are alike heck-doors ('quergeteilt'), and near together in the outer wall.

In the year 1703, Thoresby the Yorkshire antiquary, in a letter to Ray, stated that a heck was the lower half of a door : and in the Wakefield Shepherds' Play, the sheepstealer, having arrived home with the stolen sheep, calls out 'Good wyff, open the hek ! Seys thou not what I bring ?' Here apparently the lower half of the door was closed, and the 'good wyff' was peering through the opened upper half into the darkness of the night.

The word 'heck' has become 'hatch' in modern English, in accordance with custom, and the meaning has become obscured with the change. The word 'hatch' is applied to a very small door, such as buttery-hatch, and also to a door which is not hung, but is pushed along to open, a survival, probably, of the small loose, unhung door of early times. 'Heck' is now only used in dialects.

The heck or hatch door was sometimes a lattice of crossed laths, or of upright bars. Upper doors of lattice work are found in Germany, and are sometimes hung with a withy: and in 'Minne-hagen' a farmer uses a half door as a shield, and long before then the Goths used shields made of wickerwork. In the English Lake District, until recently, gates were hung with plaited withies.

In the seventeenth century, in the 'North country,' the interior of the dwelling was protected from draughts, when the half door was open, by a partition, two and a half yards high called a 'hollen.' Such screens still remain in some old cottages.

Herr Rhamm has described[1] a simple substitute for a heck-door, used in the South of Germany, in the form of a half door placed in front of a door of the full height. It is called a 'Fallter,' possibly a 'fall-door,' although it is side-hung.

In the second kind of double door the division is vertical and produces a pair of doors, meeting together at their unfixed sides. Doors of this form are known as double-hung or folding doors. The doors of the Romano-British stations on the wall in Northumberland were formed in this manner, and similar doors are still in

---

[1] *Urzeitliche Bauernhöfe*, p. 602.

use for wide openings, as the doorways of carriage houses. They are usual in old English barns, and are then invariably of some form of batten door, which is a very old type, as we have already seen.

They were also used in early times beyond the boundaries of the Roman Empire, for in the year 446 a Byzantine named Priscus went with an embassy to Attila, and afterwards recorded that the houses which were built according to the style of the Goths, were surrounded by a fence, in which was a two-leaved ('doppel-flügeliges') door[1].

Both heck and folding doors are combined in the wooden doors of the gatehouse arch at Wingfield Castle, in Suffolk, of the Tudor period[2].

The earliest fastening of the door was intended to secure it against intruders from outside, and so it was fixed on the inside of the door : and this has persisted as a custom, supported by experience, down to our own times, as modern stock and rim locks, and bands and hooks show. The most primitive of the door fastenings still in use is a loose bar which is drawn across the door and secured at the ends. Such a bar was the accepted mode of securing the door as recently as the seventeenth century in many of the stone houses then erected by the smaller gentry of South Yorkshire and Derbyshire.

In some Derbyshire churches of the Norman period, as Steetley, the doors were fastened in this manner on the interior, and a long hole was constructed in the jamb of the doorway, into which the bar was pushed when not in use. There is a corresponding, but shorter hole in the opposite jamb to receive the other end of the bar, and to hold it fast when in position, thus securing the door.

In a fifteenth century glossary, ' grapa' is translated 'as an hol to putte yn a barre,' and in the same century it is translated as the hole in which the bar rests ('est foramen in quo quiescit vectis ').

Such a mode of fastening the door by a bar was in general use in England for many centuries, but its use in the Norman period in such important buildings as churches may be taken to imply that it was then too superior a method for use in inferior, or even ordinary buildings : the difficulty of the construction of the long

[1] M. Heyne, *Deutsche Wohnungswesen*, p. 14.
[2] See illustration in *Country Life*, June 28th, 1913.

hole in the stone wall, for the bar, shows also that its original use was in buildings constructed of other materials.

There were various names in the past for the door-bar, and some of them have survived in the dialects. One of these old names was 'slot.' In the fifteenth century *Catholicon Anglicum* occurs 'a slotte, ubi a barre,' without the Latin translation, but this is given as 'pessulus' in the *Manipulus Vocabulorum*. The door-bar is still called a 'slot' in the Paisley district, where the phrase 'stake the door' is also used, with the meaning of shut the door.

'Stang' was another very early name for the bar, and the Latin word 'vectis' is so translated in glossaries from the eighth to the twelfth centuries[1], and the word has continued in dialect use. In a lecture before the Leeds and Yorkshire Architectural Society, Mr Harold E. Henderson said : ' At night an oak bar, called a stang, is fixed horizontally inside the door, with each end in a pocket cut out of the stone jambs[2],' and the word 'stang' was used in Worcestershire, in the year 1790, for 'a draw rail, or long bar, passed between two posts to serve as a gate, and drawing in and out when anyone is to pass[3].'

The bar was also called a 'spar' in the Middle Ages and to fasten the door by a bar was to spar it. In the later mediæval Wakefield play of *The Harrowing of Hell*, the command is given, 'Go spar the gates!' (against Christ), and in the *Catholicon Anglicum* 'to sperre' is translated by the Latin 'claudere, prohibere, intercludere.' In the same century the more obvious name, 'door-bar' is variously translated by 'clatrus, repagulum, vectis,' and by 'obex[4].'

It was a disadvantage of the door-bar that it could only be fixed from inside, which must have been inconvenient. In Germany an arrangement was in use by which the bar could be moved by a strap from outside the door, but no English examples are known to the writer. Mr John Ward, in his book *The Roman Era in Britain*, has illustrated ' a bar of iron, bent somewhat into the form of a sickle, with a flat handle,' of which examples are found on Roman sites in this country and in France, and also ' with Late Celtic remains in both countries, for which reason it has been

---

[1] Wright-Wülcker, *Vocabularies*.
[2] ' Minor Domestic Architecture of Yorkshire.' See *The Builder*, 23 June, 1911.
[3] *English Dialect Dictionary*, *s.v.* Stang.
[4] Wright-Wülcker, *Vocabularies*.

called the Celtic key.' He believes that it was inserted from outside through a hole in the door to move the bolt inside. Slight
variations in the position of the hole and the length of the 'key' in
each door would give security. To the writer it seems possible
that it was used to move an inner door-bar from the outside. The
writer has been informed that a method of opening the bolts of
doors similar to the so-called Celtic key is still in use in Austria-
Hungary, but the key used is a bent wire, and is of a temporary
character.

In another and better device the door-bar became the curiously
long bolt of a wooden lock, and of this, also, the writer is not aware
of the survival of any English examples. It could be locked or
unlocked from the outside, and so marks an important advance in
the development of door fastenings. Probably from the shortening
of the bolt, there were developed the wooden locks of Scotland,
and the Scandinavian countries, as illustrated by Mr Ward, and in
the *Billedbog fra Frilandsmuseet*[1].

The shortening of the bar, as in the wooden locks, was a
natural development when the hanging side of the door was
securely fixed so that it could not be opened, either by means of the
hinges, or in some other manner, and in one form the bar was only
short enough to cover the opening side of the door, and a hole was
required in only one of the jambs. At Offerton Hall, Derbyshire,
built in the year 1658, there is such a fastening, and it has an iron
ring to pull it out of the hole in the wall. It was an easy matter
to attach such a shortened door-bar to the door itself by means of
wooden staples fixed to the door: such was the ancestor of our
present forms of bolts and sliding latches. Originally it was of
wood, and the writer shows a sketch of an old wooden example
(Fig. 65) from Wigtwizzle, a remote hamlet in South Yorkshire.
This bolt was actually called a 'bar' by the tenant of the building
which it secured.

With the increase in wealth and the decrease in the cost of iron,
it became possible to make the bars of that material, which lent
itself to developments which were not possible in wood. The
principal varieties of these were the folding bar and the split bar:
in the former the bar was secured to one leaf, in the latter one half
slid into the wall, and the other half swung.

[1] For more examples of simple bolts and locks the reader is referred to an article by
Cyril G. E. Bunt, on 'Old Wooden Locks' in the *Architectural Review*, March 1915.

A further improvement was to fasten the short bolt to the door by one end. Then the bolt, instead of sliding, was used by lifting its unfixed end up and down, which was secured by some form of catch on the door frame or jamb. This form of fastening is widely known as a 'sneck,' and, reasonably following precedent, it is fixed inside the door. There are three principal methods of opening it from the outside. In the latest and most developed form, a lever passes through the door from the outside,

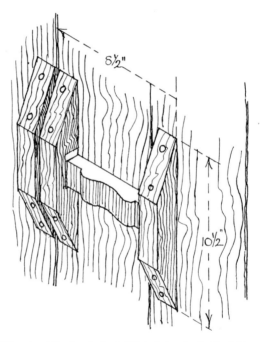

Fig. 65. Wooden bolt at Wigtwizzle, South Yorkshire, locally called a 'bar.'

under the latch on the inside of the door. The thumb latch of the present time has been reached when the lever is fixed by a hinge, instead of being loose, as in early examples.

The second method of opening a sneck is by a hole in the door, through which the finger, or a nail, or a slip of metal can be inserted to lift the latch. In such a case the latch is entirely on the inside of the door, and this seems to have been the 'haggaday' of the Middle Ages, a word which still survives in the dialects. We have seen above that this method of the hole in the door was used to open the sliding bar or lock from the outside.

The third method is to bore a small hole in the door, through which is passed a string, with its end tied to the latch inside. The latch is lifted by pulling the string. In the folk-tale of Little Red Riding Hood a bobbin was attached to the string outside. In Shropshire it is called a 'clicket,' in Yorkshire and Lancashire a 'sneck-band,' and in Cumberland a 'snecket': owing to changes, and the absence of literary standardisation, however, the meanings of these words, and others connected with door fastenings are not fixed. Whitaker says of the west wing of Samlesbury Hall, built in the year 1532, that, 'Although the building of it must almost have laid prostrate a forest, yet the inner doors are without panel or lock, and have always been opened, like those of modern cottages, with a latch and string[1].' Thus when Whitaker wrote, in the year 1818, the method of opening cottage doors was that which had been used in mansions some three hundred years before, an example of that descent in the social scale of building methods, which has been often mentioned in this book. The string method of opening doors is still resorted to in out-of-the-way places[2].

In cottages without locks, in the North, the snecks were made secure and prevented from being raised by the insertion of a piece of wood or iron in the staple. This was called a 'snib': the door was said to be 'snecked' when the latch was merely closed, and 'snibbed' when it could not be opened from the outside.

Thus, from the simple door-bar, spar or stang, may be traced, step by step, the descent of the various simple latches and locks, and probably also the framed door itself.

At the present time we look upon locks and bolts and latches as distinct, but that was not the case in the period when they were being evolved, as the contemporary glossaries show. The word 'lock' was used for the securing of a door in any manner, and even nowadays, to lock is used in the sense of to shut, and not necessarily with a key, in the dialect of English-speaking South Pembrokeshire[3].

---

[1] *History of Whalley*, p. 500.

[2] A wooden latch, worked by a string, and 'long since displaced,' and a wooden bolt or 'bar,' like that in Fig. 65, from a destroyed house at Brockhampton, Herefordshire, are shown in *Domestic Architecture in England during the Tudor Period*, II, p. 234, by Thos. Garner and Arthur Stratton.

[3] Ed. Laws, Glossary in *Little England beyond Wales*, p. 420.

The *Catholicon Anglicum* translates 'A lok : clatrus, pessulum, obex, repagulum, sera, vectis,' and in the same fifteenth-century dictionary we find 'A Barre: clatrus, pessulum, obex, repagulum, vectis,' so that all the words by which the compiler translated lock were used by him for the translation of bar, with the exception of sera. The same dictionary renders 'a snekk: obex, obecula, diminutium, et cetera, ubi a Loke.' Another old vocabulary gives 'pessulum' as 'a lytel lok of tre,' that is a little wooden lock, 'a hasp': and another defines it as 'a latche or a snecke or barre

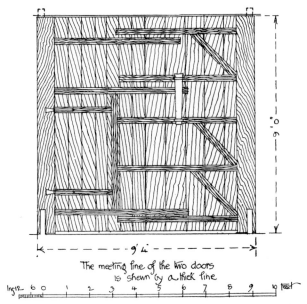

The meeting line of the two doors
is shewn by a thick line

Fig. 66. Folding doors of a barn at Wigtwizzle, South Yorkshire. The battens project on each half-door, forming a means of fastening the door. They are secured by a loose vertical bar.

of a dore.' The explanation is that the word 'lock' was synonymous with the door-bar or stang from which our modern locks are descended.

In the oldest two-leaved or double-hung barn doors of South Yorkshire the battens are prolonged on the meeting side of each door beyond the boards, and they are fixed on the inside of the door, as usual, for protection from the weather. The prolongation of the battens is for the purpose of fastening the door, and is a reminder for us that the battens themselves had their origin as a means of fastening the door. The battens are utilised in various ways, thus on a door, shown in Fig. 66, from the same remote hamlet as the fastening shown in Fig. 65, the door is secured by means of a loose wooden bar fixed vertically in iron staples attached to the battens. The drawing, Fig. 66, shows the inside of the doors, with the boards on the right half indicated by lines slanting downwards from right to left, and those on the left half by lines slanting downwards from left to right.

In other old South Yorkshire barns, the folding doors are secured by a moveable centre post. In some instances this is merely secured by insertion into a slot in the wooden lintel : in others, of a more advanced form, the head of the post is fixed in a staple, and the foot into a hole in the threshold. Such a bar, running the whole height of the door, is evidently a more primitive form of the fastening with iron staples shown in Fig. 66. In the latest and most convenient form of the latter fastening the upright bar is hinged with a peg at one end, and the other and free end fits in a staple. Thus both the horizontal bar and the vertical bar were developed into the hinged latch—the fastening of the opening side of the door.

It is not enough to secure the handle side of the door : the other side has also to be fastened. There are two principal methods of doing this in England, and both of them reach, in time, to the beginning of the historic period, that is the Roman occupation of Britain. The harr is the first method and the hinge is the second, and they are fundamentally distinct, for in the second method the door is connected to its surrounding or frame by a hinge, but in the first method no hinge or other means of attachment is required, as the door itself is directly in contact with its surroundings.

The harr is formed by prolonging the hanging stile of the door, so that its upper part suitably shaped, runs into a hole in the lintel or into a projecting 'ear,' and its lower part or a pin attached thereto, is fixed in a hole in the threshold : actually, the whole door then turns on itself and not on hinges. Such a form of door hanging has a wide historical and geographical distribution : it was used by the Ancient Egyptians, and Sir Henry Layard found that it had been employed at Nimroud. The Etruscans made the stone doors of their tombs in such a manner : the doors of a tomb at Chiusi consist of two stone slabs, and the mechanism of their hinges consists of 'two projections at the top and bottom carved out of the solid stone slabs of which they are part and parcel ; these round pivots fit into the holes in the lintel and the threshold, above and below, and the door turns on them as on hinges...in a doorway thus planned, the lintel must necessarily be superposed after the lower pivot has been adjusted into the hole in the threshold[1].'

To the Romans the harr was known as 'cardo,' and it was used by them at the stations on their wall in Northumberland. In English glossaries the word 'cardo' is translated 'harr,' with various spellings, from the eighth to the fifteenth century[2].

Gudmundsson[3] has shown that the Vikings fixed their doors in this way. It is still in use on the Continent and Gladbach, in his *Schweizer Holzstyl*, has illustrated a Swiss house with wooden window shutters hung in the same manner.

Messrs P. M. C. Kermode and W. A. Herdman show that the wooden doors of the little Manx churches known as 'keeills' were also harr-hung. At one, Lag-ny-Keeillee, the 'socket stone was in position within the doorway on the South side,' and 'just outside the building' there was found 'a large flag, 57 in. by 25 in. and about 5 in. thick ; this had a hole countersunk in its side, and it was found that this would have been exactly over the socket if set as a lintel so as to have 18 in. of its length in either side wall and to project inwards for 3 in.[4]' The authors suggest that the door was simply 'a bundle of saplings, or a gorse bush[5],' at the keeills where no socket stones were found. The only existing example of a harr-hung church door known to the writer is in North Wales at remote Llanrhwchwyn, the so-called 'Llywelyn's old church.'

The harr was undoubtedly the most common method of hanging wooden doors in England down to the end of the Middle Ages, as is shown by the number of references to it in the contemporary literature. The earliest reference is in *Beowulf*, where, in the struggle between the hero and the monster Grendel, the door of Hart Hall is said to have been burst from its harrs

---

[1] M. L. Cameron, *Old Etruria and Modern Tuscany*, p. 80.   There is an illustration of the door, from a photograph.

[2] Wright-Wülcker, *Vocabularies*.

[3] *Privatboligen paa Island i Sagatiden*, pp. 238–9.

[4] *Manks Antiquities*, 2nd edition, p. 86.   [5] Ibid. p. 91.

('Heorras tohlidene'). Hundreds of years later, Chaucer, in the prologue to his *Canterbury Tales*, described the miller as 'a lusty

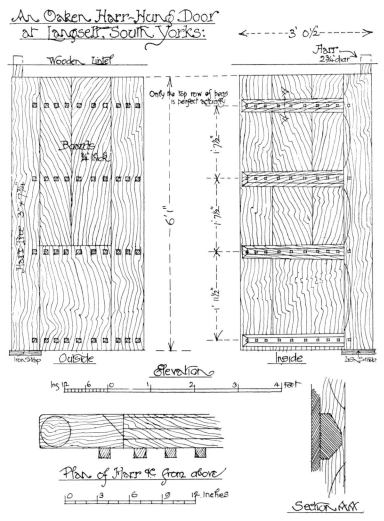

Fig. 67. The boards are pegged to the battens with wooden pegs, with chamfered heads. Such pegs served as patterns for the late mediæval iron bolts, and this is an example of the unheeding way in which one craft copied the work of another.

knarre, ther nas no dore that he nolde heve of harre.' Both Skelton and Gower use 'out of harr' figuratively, with the meaning of out of order, which shows that harrs were not only common, but also that they were often out of order.

Mr Herrtage, in his edition of the *Catholicon Anglicum*, has given a quotation from Balmon: 'The endes of this line that is named Axis, be called "cardinales coeli," and be pight in the foresaid pole, and are called "cardinales," because they move about ye hollownesse of the Poles, as the sharpe corners of a doore move in the herre.' In the Wycliffite version of the Bible, 'As a door is turned on his herre, so a slow man in his bedde,' is the rendering of the verse Proverbs xxvi. 14.

The folding doors of the barns in South Yorkshire were made with harrs until the middle of the nineteenth century; but this method did not continue in use for the smaller and lighter doors to so late a time. The writer shows drawings to scale of one of the smaller harr-hung doors of South Yorkshire, viz., the door at Langsett, which has been already mentioned (Fig. 67). This door, of a mediæval type, is now fixed in a fowl-house, but it has evidently been repaired, and is certainly not in its original position.

In a typical harr-hung single door the hanging stile, known as the harr-tree, is about 3 in. thick, and into it are framed the battens, generally three in number (but four at Langsett), nearly square in section, about $2\frac{1}{4}$ in. by 3 in., and often rudely chamfered; butt-jointed boards are nailed on these battens, flush with the harr-tree. The top of the harr-tree is cut off above the top of the boards, and worked into a round pivot of a diameter of 3 in. In some old examples an iron peg is driven into the harr-tree at its foot, which was an improvement of the earlier duplication of the harr at the top of the door: the pin turned in a hole in the threshold, and its introduction probably coincided with the use of stone for the threshold. Later the bottom of the harr-tree was clasped with an iron strap (as in Fig. 67) in order to prevent splitting.

Fig. 68 shows a sketch of the head and heel of a harr-tree from an old unused door at Falthwaite, South Yorkshire, in which the peg is apparently a 'sneck' driven into the post and protected by an iron ring as a washer. In other cases the lower peg is a part of an iron two-strap, the harr-tree being clasped by the arms of the strap.

The word 'harr' is still found in the dialects of all English counties, except the South-Eastern. In Old English the word had a masculine and a feminine form : but the word 'harr' is now applied without distinction to the peg, or the socket and to the stile itself. Its true meaning had become corrupted long ago, for in the year 1580, in the *Nomenclator*, it is defined as the back upright timber of a door or gate, by which it is hung to its post, and a few years later, in 1622, Cotgrave wrote of

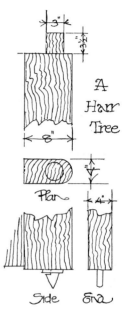

'the harr of a dore, the peece, band, or plate that runnes along on the hindge side of some dores.' The correct word 'harr-tree' is retained in some parts of the North of England, but in Northumberland the handle stile is now called the 'fore-har' and the harr-tree itself is called the 'back-har.' In the year 1800, in Yorkshire, according to the agricultural topographer John Tuke, the harr-tree was made of a harder wood than the rest of the door or gate[1].

It was necessary to fix such a door at the same time as the lintel : otherwise it might be loose and so could be 'heaved off harr,' as did Chaucer's lusty miller. In the Roman station of Cilurnum, now called Chesters, the writer has seen a door threshold in which there is a groove, worn by the pin of a badly-fitting harr-hung door.

To allow of the hanging of the door after the lintel had been fixed, a curved piece of wood, which the writer has heard called an 'ear,' was sometimes attached to the lintel, and a socket so formed to receive

Fig. 68. Door-hanging by means of harrs from Falthwaite, South Yorkshire. This was formerly the most usual method of hanging doors in England.

the upper harr. The holed stones which project from the sides of the doorways of some Romanesque stone churches in Ireland were probably fixed for some such use.

The *Dictionary of the Architectural Publication Society* says that in Roman times those who wished to enter a room unheard, took the precaution of previously throwing water upon the wooden hinges, in order to prevent them from creaking[2]. Such doors evidently had a wooden harr at the heel, and not an iron peg.

[1] *General View of the Agriculture of the North Riding of Yorkshire*, p. 100.
[2] *s.v.* Hinge.

The door-bar crossing the whole width of the door, as already described, was a useful fastening for the harr-hung door, which was never very secure against marauders.

The writer now knows of no harr-hung house door : the last survivor was removed a few years ago, and the method which was thought fit, in the fourth century, for the doors of a frontier station of Imperial Rome, is considered in the twentieth century to be too inferior for a cottage, and is only allowed to remain in barns and fowl-houses. The word 'harr' does not occur in Gothic, the oldest of Teutonic languages, and in early times only the doors of the better buildings would be harr-hung, for it is evident that if the doors in the fine stone buildings of the forts were fixed in this way in Romano-British times, then the doors of ordinary buildings would be fixed in a very inferior way, if they were fixed at all. Although the modern spring hinge usually works in a manner similar to a harr of a later kind, the harr itself followed the usual line of descent of building constructions : from the most important buildings it descended, as the centuries passed, to the less important, and then became entirely obsolete.

At the Romano-British village of Woodcuts Common, General Pitt-Rivers found circular stones with mortises sunk in, which he supposed to be for gates to turn in[1], and similar gates may yet be seen in South Yorkshire, where the foot of the back stile of the gate turns in a shallow cup cut in a stone. Sir Arthur Mitchell made an interesting observation on a similar gate. He wrote : 'I once saw the post of a field-gate, turning in a hollow in an earth-fast stone, and not one hundred yards away, I saw another gate, entirely and skilfully made of iron. The owner of the two gates thought the old-fashioned one in many respects the better, and he half convinced me that he was right. He wholly convinced me that the continued use of what we choose to call a rude mechanical arrangement is no necessary evidence of mental incapacity in the user. Both of these gates were set up by him, and he wished to know which of them was to be taken as the indication either of his capacity or his culture[2].'

A few old field-gates remain in South Yorkshire which show a method intermediate in hanging between the harr and the hinge. As field-gates could not have a lintel to turn in, a modification of

[1] *Excavations in Cranborne Chase*, I, p. 143.
[2] *Past in the Present*, p. 128.

the 'ear' was devised. In South Yorkshire the old field-gate posts are of stone; a hole was driven through them, and in it was wedged one end of an angularly-bent piece of wood called a crook ; the head of the harr-tree or hanging stile of the gate turned in the angle (Fig. 69). The English took the pivot-hung gate with them to America: its use spread westwards in the New World, but instead of being fixed with stone posts, it was balanced by weights

Fig. 69. An old method of hanging a field gate. It is a modification of the method of hanging doors by means of harrs. The example in the illustration is at Midhope, South Yorkshire.

on the top rail in a primitive manner[1], and this may be an example of the survival of methods in America after they have become obsolete in England, just as some 'Americanisms' in speech are really survivals of the standard English of earlier times.

The harr-hung door is itself in direct contact with its frame or surroundings: the second form of door hanging, the hinge, forms a

[1] L. V. C. Fay in *Country Life*, 20 Dec. 1913 (see also Ibid. 15 Nov. 1913).

connection between the door and its frame, and is also the means by which the door is rotated. In Flanders the writer has seen a form intermediate between the harr and the usual bands and hooks. The bands were of L-iron, but instead of swinging on a hook, there was a pivot forged to them which served as a harr-pin.

Of the various forms of hinge, those now known as bands and hooks were the most usual in the past. At the Romano-British village of Woodcuts Common, General Pitt-Rivers found the iron hooks for door-bands: they were used during the period of Roman wealth and civilisation, and after then, iron was hardly used as a material for the door fastenings of ordinary buildings until the sixteenth century. In Sweden the use of 'wooden hooks of natural growth' in place of hinges continued to a much later time[1]. According to John Middleton, in the year 1798, in some places in Middlesex 'the stealing of gate hooks and iron fastenings is so common, as to compel the farmer to both hang and fasten his gates with wood, which is easily done, and equally secure with iron[2].' Examples of wooden hinges in buildings are still in existence on the Continent of Europe, but the writer does not know of any that have survived in England.

The variety of the names possessed by bands and hooks is evidence for their widespread use in the past. In Kent, Surrey and Sussex they are known as 'hook and ride,' for which 1522 is the earliest date recorded in the *Oxford Dictionary*. In Lancashire, 'band and gudgeon' is used, and the word gudgeon occurs in the Nottingham Corporation Records, for the year 1496, when one was bought for the Town Hall door. The London term is 'hook and band' and for this reason it has become the standard term in use in building construction. In Yorkshire and the North of England the hooks are known as 'crooks,' and in the *Durham Household Book*[3], for the year 1530–1, three pence a pair is entered for six pairs of 'duyr bands' with 'croykks.' In the *Catholicon Anglicum*, 'A cruke of a door' is translated by 'gumphus.' In the year 1329, 'two vertinell with two gumphis' were bought for the palace at Westminster, and in the year 1405–6, the churchwardens

---

[1] Alex. Beazeley, 'Notes on Domestic Buildings in Southern Sweden' in *Transactions of the Royal Institute of British Architects*, 1882–3, pp. 117–128.

[2] *General View of the Agriculture of Middlesex*, p. 137.

[3] Published by the Surtees Society, p. 73.

of St Michael's, Bath, paid six pence for 'gomphis et vertinellis' for doors. Other terms for the 'band and hook' are 'hook and eye' and 'hook and hengle.' The earliest use of the word 'hengle' is about the year 1300, when 'vertineles' was glossed 'hengles.' Two distinct meanings of 'hengyl' are given in the *Promptorium Parvulorum*: the one 'of a dore or wyndowe' translated by 'vertebra, vectis': the other a 'gymewe,' translated by 'vertinella.' 'Verteveles' was another Mediæval Latin word for these hinges.

Joseph Moxon, the contemporary of Sir Christopher Wren, gave technical directions for hanging doors in the descriptions of joinery in his *Mechanick Exercises*. When the doors were hung with bands and hooks, the latter were to be fixed 15½ in. above the sill and as much below the top of the door[1]; battened doors were to be hung with cross garnets. Many other kinds of hinges had come into use by the early part of the eighteenth century.

The direction of development with the hinges has been to reduce their size and to place them out of sight. Our modern butt hinges are descended from the so-called cock's head hinges of the Renaissance, and this was a decorative improvement upon the earlier hinges and long bands. Such hinges as the cock's head, in which the portion fixed on the door was like that on the frame, were called variously 'jimmers' or 'gimmers,' 'gemows,' and 'gemmels,' all of which are derived from the French *jumeaux*, twins, because they seemed like twins to the mediæval wright. The earliest record for gemow in the *Oxford Dictionary*, is from the *Memorials of Ripon*, for the year 1396. There was the same confusion in the names of hinges as in other sections of building construction which had not become standardised through literary use and the continuous changes and improvements caused further confusion.

Butt hinges are now generally used, except for special purposes. The old string or band which was used, as we have seen, in Dr Whitaker's time, to open the doors of cottages, is no longer used, and the 'sneck' or latch, which followed the band, is now almost as obsolete as the band. Its place has been taken by the knob, either alone or in union with a mortice lock, for the old stock lock, like the old sneck, is now only used in inferior situations.

[1] *Mechanick Exercises*, p. 160.

CHAPTER XV

WINDOWS AND CHIMNEYS

The combined light- and smoke-hole—Perforated slabs—Shutters and windows
—Lattices—Sash and casement windows—The chimney and its descent—
The louvre and the lantern—The smoke chamber and the flue.

The opening in the building for entrance and exit, the doorway,
has now been described and there remain the openings for the
admission of light and air and the removal of smoke. On the
Continent there is an abundance of material in existing buildings
for the study of primitive forms of these openings[1], but in
England, where improved methods have everywhere superseded
the old, the enquirer is almost wholly restricted to references
in literature, and to entries in building accounts, so that our
knowledge of the lighting and of the disposal of the smoke in
ordinary buildings is only scanty before comparatively recent
times.

Although a building must obviously have a doorway of some
kind, it is not necessary for it to be provided with openings for the
admission of light and for the outlet of smoke.

The so-called 'black houses' in the island of Lewis, described
by Sir Arthur Mitchell, had neither window nor chimney : there
was not even a hole in the wall. The absence of a window in
the houses of Lewis was very general, and in the year 1880 Sir
Arthur Mitchell wrote that he had seen thousands of similar
houses in various parts of Scotland. Such light as they obtained
came through the door or through one or two small holes in the
eaves of the roof at the top of the wall, or through chinks due to
deficiencies in the construction of the roof. Not only was there

---

[1] As an instance some thirty-five pages were required for the description of the light-
and smoke-hole ('lichtloch' and 'rauchloch') in Teutonic buildings, by Herr K. Rhamm,
in his *Urzeitliche Bauernhofe*.

no chimney, but there was not even a smoke-hole, so that the inside of the house was in a constant cloud of peat-reek, and such smoke as was not deposited on the thatch oozed out over the whole roof, giving to the house 'the general appearance of a dung-heap in warm, wet weather[1].'

A more recent description of the 'black houses' or 'clachans' says that 'generally in the huts, light is admitted by small apertures in the roof. In some cases these are glass-covered, the single panes of glass being simply let into the straw without sash or woodwork of any kind, and the art of glazing is thus reduced to its utmost simplicity. In others they are just holes made in the straw, like a rabbit-run in gorse, and could easily be blocked to keep out the wind by a wisp of grass or straw.' There are no chimneys: the smoke finds its way out through the roof, and the atmosphere is indescribably dense and impure[2].

Sir Arthur Mitchell's object was to describe survivals of once common methods and practices which had become obsolete, and we have seen, in earlier chapters, that survivals are to be sought in the North and West. We may therefore conclude that the window-less and chimneyless 'clachans' of Lewis are the survivals of methods of building which had once a wider distribution.

There is no window in the South Yorkshire charcoal burner's hut, and the only light is through the small doorway, while in the hut of a more advanced type, from the South of England, mentioned in chapter IV (p. 25), the 'heck-door' is used for the admission of light and the hut is quite dark when the door is closed. It is probable that this form of door was the result of an attempt to use the doorway for light and security at the same time.

The earliest windows were simply holes in the walls or roof, and not protected in any way, so, in the *Ancren Riwle*, the window is called a 'thurle,' that is a hole[3]. Such holes survived into the nineteenth century: each of a pair of cottages or huts built 'on crucks' in Berkshire, described by Dr Guest in his *Origines Celticae*, was lighted only by a hole in the gable, for the eaves were so low as barely to admit of doors, and the same holes evidently served to let out the smoke. In some of the smaller

---

[1] *Past in the Present*, p. 53.

[2] H. Whiteside Williams, *The Reliquary*, VI (1900), p. 76.

[3] Mediæval examples of holes used as windows were cited by Mr Joseph Wright, in his *History of Domestic Manners and Sentiments*.

houses of North Wales mere barn slits 'in the wall,' serving as windows, were sometimes used in the back walls of the 'chamber' as late as the seventeenth or eighteenth century[1].

Professor J. H. Middleton has written of the eleventh-century 'aula regia,' or 'chapel' at Deerhurst, that 'In one window what appears to be part of its original oak casement still remains firmly built into the masonry of the wall; it is a simple slab of oak, two inches thick, in which a small round-arched opening has been cut. The edge of this opening has no trace of any filling-in, and the window was probably open to the air[2].' If such a window, 'simple' as it is, was to be found in an 'aula regia,' literally a royal hall, then the ordinary buildings of the period must have had fittings which were still more 'simple.'

In certain churches of Norfolk there are double-splayed round windows of pre-Norman date, as for example, at Witton-by-Walsham Church, where two openings remain which form part of a range of upper unglazed windows, of an early church of the tenth or late ninth century. They are about 15 ft. from the ground and only 6 in. or 7 in. in diameter, with an outer diameter of the circular splay of about 18 in.[3] The inconvenience, if not the danger, of such unprotected windows is obvious, and that must have been greater when they were in the slope of the roof.

The Anglo-Saxon word for window was 'eag-thyrl,' that is eye-hole, and it may have meant either a round hole like an eye or more probably a hole to which a person applied his eye, when he wished to look out; the form and size of the contemporary windows make either derivation possible. Our modern word window is derived from the Old Norse word 'wind-auga,' that is a wind-eye. The Scottish word 'winnock' is a less changed form of the original word, and it is an indication of the small size of early window openings that this word 'winnock,' according to the *Dictionary of the Architectural Publication Society*, is used as a diminutive. The present Swedish form of the word, 'vindoga,' is used for an opening in the roof. Another Anglo-Saxon term for window was 'wind-dur,' that is wind-door.

The window openings were closed, from early times, in various ways, very usually by a fixed wooden slab, or by a movable

---

[1] Harold Hughes and H. L. North, *Old Cottages of Snowdonia*, p. 5.
[2] 'On a Saxon chapel at Deerhurst, Gloucestershire' in *Archæologia*, L, p. 68.
[3] Handbook by Rev. Dr J. C. Cox, I, p. 254.

wooden board or shutter. The remains of wooden perforated mid-wall slabs which served as windows have been found embedded in the centre of the walls of English churches built before the Norman Conquest. The slabs had been walled in at a later period, when such primitive methods of lighting had become too old-fashioned to be retained. Similar oaken wall slabs perforated with round holes still remain in the circular lights to the West of the so-called 'annex' at St Peter's Church, Barton-on-Humber[1]. The apertures in the boards do not usually form an ornamental pattern, but in the scanty remains of such a wooden slab, found in the north wall of the chancel of Birstall Church, near Leicester, in the year 1869[2], the piercings were designed in a twist-and-ring pattern, of a type usually associated with the Anglo-Danish period.

At Earl's Barton Church, the twist-and-ring pattern is perforated in a stone window-slab. These eleventh-century slabs, pierced with the type of plaitwork patterns fashionable at the time, did not continue in use and had no further development. The pierced slabs of English window openings, whether of wood or stone, had their origin in the pierced marble slabs with which window openings were filled in the Mediterranean lands, and the eleventh century English builders, in making use of the plaitwork patterns took advantage of the large amount of background in the patterns of the period.

The modern descendants of the plainly perforated pre-Conquest mid-wall slabs are the boards pierced with holes, generally round, and arranged in a pattern, which are still used to fill ventilation openings in pantries and similar rooms of old farmhouses in South Yorkshire and elsewhere.

A window opening of the pre-Conquest chancel of Jarrow Church is filled with a stone slab, pierced with a tiny round hole, evidently the 'eag-thyrl' of Anglo-Saxon speech. (See Fig. 70.) It has been claimed that this round-holed slab filling was put in as a protection against Danish raiders. The illustration also shows the walling, copied from the Romano-British manner of the North, without brick lacing courses, and very probably built of old material[3]. We have seen that the slab in the window opening at Deerhurst 'chapel' had a small round-arched opening cut in it. In the twelfth century the designers of churches combined the piercing

[1] W. F. Rawnsley, *Highways and Byeways in Lincolnshire*, p. 101.
[2] Baldwin Brown, *Arts in Early England*, II, p. 203 et seq.   [3] Ante, p. 148.

of round holes and of round-arched openings, in stonework, under one containing arch, and from such a combination of the two early forms there sprang a development which culminated in the splendid window tracery of the Middle Ages. Such great windows as those of the choir of Gloucester Cathedral are the direct descendants of tiny windows like those of Deerhurst 'Chapel' and Jarrow Church, by way of 'plate' tracery, which is obviously of wooden origin.

Fig. 70.   Chancel wall of Jarrow Church, Durham.   To
the right is a pre-Conquest or 'Saxon' window
filled with a stone pierced with a small round hole.
This is the 'eag-thyrl' (lit. eye-hole) which served
as a window at that period.

In the course of time, with cultural progress, the window opening was gradually made larger : it was then wasteful to cut such a frame as that at Deerhurst out of a slab, and the window frame was made of separate pieces of wood framed together. The earliest wooden frames of the kind, known to the writer, are of the thirteenth century at Lincoln Cathedral. Here again, we may apply the rule that new constructions first appear in the most important buildings, and conclude that if wooden window frames were used in a thirteenth-century cathedral, something much inferior must have been used in the ordinary buildings of the time, if they possessed windows at all.

In closing the window by a fixed slab it was a disadvantage that the amount of light and air could not be regulated, and this was avoided by the use of a movable board or 'shutter.' In Bede's *Ecclesiastical History*, one of king Edwin's chief men compared the life of man with a bird, flying through the hall on a winter's night, and out again into the darkness[1]. Moritz Heyne considered that the 'doors,' through which the bird is said to have flown, were window openings closed by wooden shutters[2], and for centuries the term 'window' was applied to the wooden shutter. In the time of Henry III such shutters were called wooden windows ('fenestris ligneis'), and the *Catholicon Anglicum* two centuries later translated 'asser' as 'a burde, siche as dores and wyndows be made of.' In the *Durham Household Book*[3], in the year 1533-4, 'dowrebands and wyndowbands' were bought together : here the windows were like our modern shutters. Gladbach, in his *Schweizer Holzstyl*, has illustrated a Swiss building with its window openings closed by wooden shutters, and the external hinged shutters so usual in the older houses of Sheffield are the descendants of the 'wyndows' of the Middle Ages. As recently as the year 1661, Ray found that only the upper part of the window was glazed in the royal palaces of Scotland ; the lower portion had merely two wooden shutters. In the so-called house of Memling, at Bruges, the highest window is closed with battened doors for shutters, in which small glass panes are inserted.

The windows, or as we should call them, the window shutters, were similar in construction to the doors, and T. Hudson Turner thought that they were very coarsely made in the Middle Ages[4]. Both shutters and doors were fastened in a similar manner. The *Promptorium Parvulorum* speaks of 'Loke, sperynge of a dore or wyndow,' and we have seen that the 'sperynge' was the placing of a 'spar' or bar across the door or window to secure it. In important buildings the windows, like the doors, were secured with bars of iron, instead of wood, as at Coggs, Oxfordshire[5].

---

[1] Bk ii, chap. xiii.   [2] *Deutsche Wohnungswesen*, p. 29.
[3] Published by the Surtees Society, p. 258.
[4] *Domestic Architecture in England from the Conquest to the end of the Thirteenth Century*, p. 82.   [5] Ibid. p. 80.

The shutters were not always hung at their sides. The wooden shutters most frequently represented in the illuminations of the thirteenth and fourteenth centuries were attached to the window head outside and pushed up and kept raised by a prop of wood or iron[1]. According to Mr G. T. Clark the embrasures of the castles of Alnwick and Conisborough contained hanging shutters which were fixed like a modern roller towel[2], and passages in the sagas seem to indicate that doors were hung in the same manner. A theory has been advanced that certain perforations in the projecting lintel of a church built of stones put together without mortar at Kilmalkedar, were for the fastening of a door hung in this manner[3] : but it has already been suggested that these were holes for 'harrs.' The present-day descendants of the shutters hung at the top are to be found in old cutlers' workshops in Sheffield, of which the window openings are unglazed, and closed by wooden shutters when not in use. Sometimes these are fixed inside the building, and hung at the top, in which case it is necessary to lift them when light is required and to secure them by a hook from the ceiling: more usually they are fixed outside, hung at the bottom, and lowered when light is wanted in the workshop.

The accounts of the churchwardens of St Michael's, Bath, are very perfect and commence in the year 1349[4]. The parish, as stated in a former chapter, possessed house property, and for two hundred years the churchwardens' accounts contain numerous references to doors and windows : the accounts show that both were the work of the carpenter, both were made of wood, and both were fastened with 'twystes and hokes.' We find that glazing was done to the windows of the church, but not to the windows of the houses, and in the financial year 1535-6 the churchwardens paid eight pence for one hundred laths to make a window for the church tower, but there is no entry for laths for the windows of the houses : it may be supposed, therefore, that the 'windows' of the houses were simply wooden doors, or as we should now call them, shutters.

[1] *Domestic Architecture in England from the Conquest to the end of the Thirteenth Century*, p. 82.
[2] *Mediaeval Military Architecture in England*, 1, pp. 181 and 446.
[3] Dr Joseph Anderson in *Scotland in Early Christian Times*, 1, pp. 107-8.
[4] These accounts have already been referred to so often, that it is hardly necessary to say that they were printed by the Somersetshire Archæological and Natural History Society in its *Proceedings*, 1877-80.

The boarded shutters were continued in use, for the purpose of security, after the adoption of glazed windows. Thus, a modern window protected with shutters, is, in one sense, a compound window of two kinds. The shutters followed the usual path; from the important buildings, such as the royal palaces, their use descended to the ordinary buildings, such as cottage houses in Sheffield, and then they became obsolete. They also followed the usual development in the direction of complexity, and their position was changed also from the outside to the inside of the glass. Professor F. M. Simpson has described an early form of the internal shutter, which consisted of two hinged flaps, one of which lay against the jamb of the window, and the other folded back against the inside of the wall[1]. The writer has found a seventeenth-century example at Onesacre Hall, in South Yorkshire, where the shutters are merely boards and protect stone-mullioned windows. Shutters of this kind were developed into the boxed shutters, in which the panelled leaves of the shutter, of slightly varying widths, are folded on each other, into a jamb recess or 'boxing.'

Although there is no evidence for the use of latticed windows in the houses in the Bath accounts, they were in very general use at the time. In Mr Ould's book on the cottages of the Western Midlands, there is an illustration of the picturesque falcon-house at Bullers, near Weobley[2]. It has a wattle ventilation panel in the gable, which suggests that windows may have one line of descent from wattlework, left unplastered, as wattled dairies are left at the present day in Albania.

According to Professor J. H. Middleton, a row of holes in some round-headed windows of pre-Norman date, found in Avebury Church, contained the stumps of willow twigs, showing that they had once been filled-in with a wattle screen[3].

The word 'lattice' is common in English literature, which is good evidence for its former prevalence. Lattices were formed of bars or stanchions crossed diagonally and framed into heads and sills: sometimes they were merely formed of crossed laths, or, as we have seen, of wattle. They were either unglazed, and closed with a shutter, or the lattice work was filled ' with thin horn or tile or

---

[1] *Building Construction*, II, p. 305, in the Architects' Library.

[2] J. Parkinson and E. A. Ould, *Old Cottages, Farmhouses and other half-timber buildings in Shropshire, Herefordshire and Cheshire*, Plate LXV.

[3] 'On a Saxon chapel at Deerhurst, Gloucestershire' in *Archæologia*, L, p. 68.

with canvas[1].' Wm Horman in the year 1519 wrote 'I wyll
haue a latesse before the glasse for brekynge[2].' They seem to
have been used by the Romans, and Messrs Hughes and North
say that the wooden lattice, in the mountainous district of North
Wales, 'did not entirely go out of use in the cottages till the end
of the seventeenth century,' and that it can still be seen in old
outhouses. In North Wales the frame is sometimes of one, but
more usually of two lights; occasionally there are three lights and
in such a case there is a touch of English influence. The frame is
closed with solid wooden shutters behind, opening inwards, and the
lattice is of a diamond pattern, which keeps out the rain better
than the square pattern, which is very occasionally to be found in
examples of the eighteenth century[3]. The lattice was also called
trellis, and it is so called in the *Durham Household Book*[4] in
1532–3. In the year 1565 'clathare' was translated by 'to close
with cross bars or trellis, to lattice up.' The lattices were not
satisfactory and in the year 1562 it was written that 'lattice
keepeth out the light and letteth in the winde.'

The lattices of the windows of public-houses were painted red[5],
perhaps with that decoction of bullock's blood already described.

Early lattice windows were probably simply of upright twigs
or branches, but it was found that crossed diagonal bars kept the
rain out better. When lead-glazed windows took the place of the
open lattice, the diagonal crossing of the lattice, according to
established custom, was followed for the lead cames, and this is the
origin of the common diamond-shaped panes. No example
of ancient glazing in lattice work is known to the writer to have
survived: it is even doubtful whether such a form ever existed,
but there are countless examples of its derivative, in which
the glass is in diamond-shaped panes formed by lead cames.
Leaded lights are draughty, liable to let in the rain, and
frequently in need of repair: they are, therefore, but little used
in living rooms, and their use survives in situations where comfort
is not of great importance, as in staircases and entrances, and in

---

[1] J. H. Parker, *Domestic Architecture in England from Edward I to Richard II*,
p. 38.
[2] *Vulgaria*, p. 242.
[3] *Old Cottages of Snowdonia*, p. 15.
[4] Published by the Surtees Society, p. 163.
[5] *Notes and Queries*, 9th Series, VIII, p. 234.

those buildings, such as churches, where it is thought desirable
to preserve a feeling of the old tradition.

Although rectangular panes are much cheaper, it was long
before the glazier abandoned the traditional shape. Generation
after generation the master craftsman taught carefully and minutely
to his apprentice the traditional method of doing the work of the
craft. When however the craftsman was called upon to copy the
work of another craft in another material, he copied, almost blindly,
the design used in the other craft, whether it was suitable to the
new material or not. The plumber copied in lead the wooden
lattice of the carpenter, and the mason copied the newel-tree of
the same craftsman's wooden staircase, laboriously working a copy
of a bit of the newel-tree on each stone step. To-day it is a vogue
to deplore the decline of tradition in craftsmanship, but tradition
is not altogether a good thing, for if the appropriate use of
materials is good Art, tradition may lead in quite another
direction.

Randle Holme in his *Academy of Armory*, published in the year
1688, gives information as to the practice of lead-glazing in his
time : he says that the diamond-shaped pane 'set Arras like with
the point upwards' was called 'A Quarry,' and a pane 'flat side
upwards' was called 'A Square.'

The use of glass for filling the window frames of ordinary
buildings is comparatively recent. Early windows, not only
in buildings of the poorer sort, were often filled with woven
material. Some kind of canvas, probably oiled, was used for the
chapter-house windows of Westminster Abbey in the year 1253:
and in the *Ancren Riwle*, the 'thurles' or small window openings
are spoken of as being covered with a 'thurl cloth.' Mr S. O. Addy
has given an instance[1], from the last century, of the windows of a
house hung with harden, a kind of coarse cloth, and the material
is still used for filling-in the window frames of outhouses. The
Sheffield workshops in the middle of the eighteenth century are
said to have been 'starvation places, mud floors in the bottom
rooms, and loose open boards to the chamber, with a ladder and
trap door, while the top was open to the moss stuck slates : no
glass in the windows, in winter oiled paper[2].'

According to Aubrey, an antiquarian writer of the seventeenth

[1] *Evolution of the English House*, p. 41.
[2] *Sheffield Iris*, 29 June, 1830.

century, glass windows were rare, except in churches and gentlemen's houses, before the reign of Henry VIII, and to his own
remembrance, copyholders and poor people in Herefordshire,
Monmouthshire and Shropshire had none before the Civil War,
which is an example of the survival of a less advanced civilisation
in the West. Harrison, the Elizabethan topographer, said much
the same, and Wm Horman, early in the sixteenth century, wrote
that 'Glazen wyndowes let in the lyght and kepe out the winde.

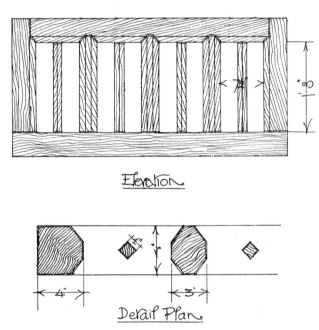

Elevation

Detail Plan

Fig. 71. An unglazed window at Falthwaite, South Yorkshire. It is
entirely of oak, and the joints are shown by double lines.

Paper or lyn clothe straked across with losyngz make fenestrals
instede of glazen wyndowes[1].' In the year 1505 it was held that
windows belong to the heir, and glass to the executors, 'because
the house is perfect without the glass': but in the year 1599 it
was decided in the Court of Common Pleas that the glass could
not be removed as 'Without glass it is no perfect house[2].' Nine
years earlier Alderman Robert Birks, of Doncaster, left by will, to
his son Robert, all 'the seeling work and portalls' in and about

[1] *Vulgaria*, p. 242.   [2] *Notes and Queries*, 1st Series, IV, p. 99.

the house where he dwelt, 'with all doors, glass windows,' etc., and then he bequeathed his house to his wife for her life[1].

Glazed windows became usual in ordinary houses in the seventeenth century, and the windows with stone mullions, common in the stone-producing districts, were copied from their unglazed wooden predecessors. Such an unglazed window of oak is shown in Fig. 71, from a farm near Falthwaite, South Yorkshire. It has been built into a wall of later date, and is now mutilated, and blocked up internally with stone or 'grey' slates. The height of each light is double the width, and the bars which protect the openings are upright, one inch square in section, set diagonally and with the angle outwards. Although this window seems very primitive in comparison with our double-hung sashes, it is elaborate when compared in the other direction, backwards, with its prototype, the hole pierced through a board.

It is evident that it was never intended to be glazed, and a somewhat similar window in a house which Mr Lutyens has re-erected for a client at Great Dexter, Sussex, was protected by a shutter sliding in grooves, which still remained, and has been copied by Mr Lutyens for other windows in the house[2].

Mr Ralph Nevill says that open unglazed windows were usual enough in the older Surrey cottages, but not often found still remaining, except in barns and stables. They were formed by stanchions set diagonally and framed into head and sill. The opening was closed with a shutter[3]. In the interesting Elizabethan mansion, named Plas Mawr, at Conway, there were unglazed windows, on a larger scale than those in Surrey and South Yorkshire. They were made with stanchions as in Surrey, but the sides of the stanchions, or mullions, were four inches wide. Similar frames were used for screens, and are an example of the absence of specialisation in early forms of construction. A very early example of screenwork, of the thirteenth century, at Exeter Cathedral is formed like the Surrey windows, with plain bars set diagonally.

Windows like these were the models for the square-headed glazed windows, with chamfered and filletted mullions and jambs, of the stone and brick houses of the Early Renaissance. But stone-mullioned windows had been used long before, and an example of

---

[1] *Notes and Queries*, 1st Series, IV, p. 328.
[2] *Country Life*, January 4th, 1913.
[3] *Old Cottage and Domestic Architecture in South-West Surrey*, p. 12.

the Romano-British period, discovered some time ago at Corbridge
(Corstopitum), has no fillet on the edge of the mullion, which shows
its derivation from wooden bars similar to those of the unglazed
windows described above. The fillet to protect the arris at the
meeting of the chamfers is a necessity in stone, but not in oak, and
the chamfers themselves are a more natural piece of craftsmanship
in wood than they are in stone.

After a time the chamfering of the stone jambs and mullions
was abandoned in South Yorkshire, and the edges were left square,
and in the eighteenth century the wooden frame again became
fashionable, an example of a rather curious retrogression and of a
certain aimlessness in the movement of modern architecture.

The glazed frames and the mullioned windows presented the
craftsman with difficulties in devising means for their easy opening.
Sometimes in the fourteenth century, as at Meare, in Somerset,
the glazing was fixed in wooden frames or casements, which were
movable, and stored when the family was away[1], but when glass
became cheaper such a method went out of fashion. It may be
incidentally remarked that the windows at Meare were like those
of the Royal Palaces of Scotland, already mentioned, in that the
upper part only was glazed, and there were shutters below, which
were merely battened doors[2]. This was not practicable in small
houses: other methods, such as iron casements, were too costly,
and so their windows were not made to open. Mr W. F. Price,
writing of the homes of the yeomen and the peasantry in *Memorials
of Old Lancashire* says that the windows were usually very small,
ventilation being provided by the ever open door : and this was
the case not only in Lancashire but all over England.

Moxon in his *Mechanick Exercises* explained the construction
of the wooden windows which were in use in his time. He says
that windows have head-sills, ground-sills, and 'transums.' In brick
houses it was usual for their 'Tennants' to run through the outer
'Jaums,' about four inches beyond them, and they were buried
in brickwork. In a timber house the head and sill were sometimes
framed into the posts, and there were no separate window jambs
or posts, 'but the better way is to frame a Window as a Brickwork
Window, and to project it an Inch and an half beyond the side of

[1] J. H. Parker, *Domestic Architecture in England from Edward I to Richard II*,
p. 36.
[2] *Dictionary of the Architectural Publication Society*, *s.v.* Window.

the Building, and to plaister against its sides, for the better securing the rest of the Carcass from the weather[1].'

In early use, glazed frames of wood, as distinguished from leaded windows, were called sashes, and so the novelist Richardson, in *Clarissa*, in the year 1747, wrote of a sash-door, which was a door fitted with a glazed sash in the upper part.

Sashes and casements, which are the forms of windows in general use at the present time, are descended from unfixed frames of the old type of window with mullions. The English word 'sash' is derived from the French word 'chassis,' and 'shas' and 'shash' are early spellings of the word in English, which continued in use until the middle of the eighteenth century. Moxon, in the year 1700, in his *Mechanick Exercises*, mentions 'shas frames and shas lights.' When wooden casements or sashes were made to open, at first they were constructed to slide, either horizontally or vertically. The earliest sashes were unweighted, and were raised by hand : perhaps it was of such that William Horman wrote in the year 1519, 'I haue many prety wyndowes shette with louys goynge up and downe[2].' For long it was usual to make one sash only—usually the upper—movable, and such windows were used in Derbyshire, in houses built during the nineteenth century. In the early examples the under sash, when raised, was kept at various heights by means of a series of notches and a catch to hook into them.

Our modern method of hanging sashes by a weight and line moving over a pulley, dates from the latter half of the seventeenth century, and is supposed to have been acquired from the Dutch, as were many other improvements at this period. In very early specimens the woodwork was massive and clumsy, and the whole of the frame, including the groove for the weight, was worked out of the solid[3].

The French seem to have derived their knowledge of sash windows from us, for the owner of a house, near Montmartre, which was being built in the year 1699, showed his sash windows to some English travellers and pointed out 'how easily they might be lifted up and down, and stood at any height : which

---

[1] *Mechanick Exercises*, Joinery, p. 148.

[2] *Vulgaria*, p. 244.

[3] If an illustration in the *Builder*, v (1847), p. 279, showing a sash frame discovered at Wickham Court, is to be trusted.

contrivance he said he had out of England, by a small model brought on purpose from thence : there being nothing of this *poise* in France before[1].' In this country the use of sash windows became general in buildings of the better sort, in the reign of the Dutch King, William III, and in the year 1688, the *London Gazette* contained an advertisement which offered 'glasses for sash windows' for sale.   Their introduction into the provinces was gradual.   Tong Hall, built in the year 1702, is an early instance of their use in Yorkshire, and Darnall Hall, near Sheffield, built in the year 1723, has sash windows, single hung, with very thick bars to the sashes, but otherwise fully developed. A writer in *Notes and Queries*, said that the first sash windows in the Swalfield district of Oxfordshire, were fixed in a house at Sileford-Gower, in the year 1728[2]; and according to another writer in the same journal, it was not until a generation later, that the first sash windows were used at Wymondham, in Norfolk.

Until the end of the eighteenth century, the sash-frames were fixed in openings without rebates.   Previous to a statute of Queen Anne's reign, the windows were fixed flush with the outside face of the wall, or only set back about two inches.   The statute enacted that in London, after 1 June, 1709, 'no door-frame or window-frame of wood to be fixed in any house or building, within the cities of London, etc., shall be set nearer to the outside face of the wall than four inches[1].'   It continued in use in London until quite recently.   This law, in which we may perhaps see the result of some observation during the Great Fire of London, did not affect country districts, where such frames continued to be set flush with the face of the wall until the adoption of the Model Bye-Laws.

In very early sashes, the bars were not mitred, but made with a mason's joint, bringing about square blocks at the corners[1]. In such of these early examples as remain unaltered the bars are very thick as the joiner was still under the influence of the old mullions.   The bars were made thinner about the end of the eighteenth century, and, still later, they were made unpleasantly thin.   Probably this was done for several reasons : there was the desire for 'prospects' in the age of romanticism : people wished to look out of their windows, as well as to have their rooms well lighted, and it was found also that the thick bars added to the

[1] *Dictionary of the Architectural Publication Society, s.v.* Sash.
[2] 3rd Series, VII, p. 509.

weight of a window which had to be lifted whenever it was opened. In the year 1840, a machine for making sash-bars, which had been invented by Sir Joseph Paxton, was premiated at the Society of Arts. It was first extensively used for the sash-bars of the Great Exhibition of 1851[1].

In the days when the top sash was a fixture, 'the lower one was usually secured by a pair of brass hooks fixed to the side beads, which hitched on to the meeting bar when closed : both had to be lifted up simultaneously before the sash could be raised[2].' The use of sash fasteners fixed on the meeting rails inside, began about the year 1776: the barrel sash fastener, with screw and spring, was an old and favourite form. Since then one patent sash fastener has followed another in a hurried succession, and probably no other fitting in joinery has so exercised the ingenuity of the joiner of an inventive turn of mind. A sash fastener is a small object and attempts at its improvement call for no mental strain, and the making of the models involves no expense beyond the inventor's means. The growing use of sash cords also called out inventiveness, but in a lesser degree, and an early patent is dated 1774.

Sash windows became popular and the fine mullioned windows in many an old house were hacked out and replaced by sashes, too often thin and inappropriate intruders on the more substantial work of an earlier time. Their use was general in new buildings also, until the irregularly named 'Arts and Crafts Movement' came in with its revival of old methods and styles and the casement window was given its old place in design. But the present is an age of rapid changes and the Georgian revival has brought sash windows into fashion again for buildings of the style. Even in this century of rapid change, the fashion of casement windows has not reached some parts of the country, and a few years ago some agricultural labourers refused to live in cottages fitted with casement windows which had been built by a certain District Council. Casement windows are awkward to open, difficult to clean, and hard to make weatherproof without incurring a certain amount of extra expense, but the labourers objected to them as tokens of an inferior social status, and in this we have a warning of the difficulty in dating old ordinary buildings.

[1] *Dictionary of the Architectural Publication Society*, *s.v.* Sash-bar.
[2] Ibid. *s.v.* Sash.

Recent improvements in windows, both sash and casement, have generally had the easier and safer cleansing of the outside of the windows as their motive.

Although there is an abundance of material for the history of the window proper in English ordinary buildings, there is hardly any for the study of the opening which served in early times for the double purpose of letting in light and as an outlet for smoke. This opening became specialised as the window when it was situated in the wall, and when in the roof, it was developed in one direction into the roof-light, louvre, or skylight, and in another direction into the modern chimney.

We know that at first no special provision was made for the departure of the smoke, but that it was left to get away through the doorway or through chinks in the roof, as it does to-day in the charcoal burners' huts, and in the 'black houses' of Lewis, and as it did also, apparently, in the poor widow's cottage in Chaucer's tale of Chanticleer and Pertelote, where we are told 'ful sooty was hir bour and eek hire halle[1].'

The next development was to make a hole in the roof, or in the wall, which served as both a light-hole and a smoke-hole. Such a hole is common on the Continent and its use is of great antiquity. According to K. Rhamm, the Old Bavarian *liehe*, the light- and smoke-hole, was a rectangular hole in the wall, over the door[2]. Rhamm supposed that the existing examples were descended from it, and these are closed by a board ('Schieber'), sliding horizontally, and moved by a jointed handle ('Hebelstange'). These holes were in the wall, but in this country they were usually in the roof, as they were in Scandinavia. Writing of the houses of the peasantry of the Isle of Man, in the year 1812, Thomas Quayle, himself a Manxman, said : 'Chimney there is none, but a perforation in the roof, a little elevated, at one end, emits a great part of the smoke from the fire underneath : the embers burn on a stone placed on a hearth without range or chimney : the turf smoke, wandering at random, darkens every article of furniture[3].' Probably the Manx 'perforation' had its origin in a light-hole. In the Irish 'cabins' sixty years ago, according to T. L. Donaldson, the smoke-vent was frequently a mere aperture left in the thatch, lined by small

---

[1] *Canterbury Tales. The Nun's Priest's Tale*, line 12.
[2] *Urzeitliche Bauernhofe*, pp. 340–5.
[3] *General View of the Agriculture of the Isle of Man*, p. 22.

pieces of board[1]. One line of descent of our modern chimneys is from 'elevated perforations' similar to those of the Manx cottages.

When the opening was in the roof, it let in the weather, and for its protection there was devised a smaller raised roof. Such a covering has a great antiquity on the Continent, and in Mediæval Latin it was called a 'testudo.' As Aelfric thought that the word meant a boarded roof, and so translated it, it may be argued—not on very sufficient grounds—that the raised roof was unknown in his time. From this 'testudo' has come the raised louvre, or lantern, of the halls of university colleges. In the Middle Ages all the light- and smoke-holes seem to have been called 'louvres,' and as the word has remained in use with their later developments, it has now various meanings, such as in the generally used term 'louvre boards.' In the dialect of Craven, the chimney is called a 'louvre.' John de Garlande, a writer of the first half of the thirteenth century, said that the light entered the house through a 'lodium,' and more than two centuries later the *Catholicon Anglicum* translated the word by louvre. Baret, later still, had 'loouer, or tunnell in the roofe, or top of a great hall to auoid smoke, fumarium.' It is unlikely that ordinary buildings in the Middle Ages had any special arrangement for the disposal of smoke, and the state of things then existing survives in the 'Black Houses' of Lewis[2].

The introduction of the vertical gable end was followed by the placing of the hearth at the end of the building instead of the middle of the floor, as in early structures. Although the smoke then went out at a hole in the roof adjoining the gable, the atmosphere must have remained unpleasantly smoky as in the cottage already cited from the *Satires* of the Elizabethan Bishop Hall, where the opening was formed of 'a headless barrel,' and the cottage was very sooty, 'a whole inch thick, shining like blackmoor's brows[3].' A further stage was the making of a sort of smoke chamber occupying all the roof. This was, indeed, a natural development reached by ceiling the building at the level of the wall-plate, and such smoke chambers still remain in the Continental countries. With the contraction of this smoke chamber to the portion of the building immediately over the fireplace, there was obtained that capacious form of chimney or flue, which was in

[1] Royal Institute of British Architects *Sessional Papers*, 1857–58, pp. 145–154.
[2] Ante, p. 248. [3] Bk v, S. 1.

use in England for so long a time.   It has already been described
as springing from the fireplace beam and the two cross beams or
trimmers, which ran into it, and in Fig. 72 there is shown the
remains of such a chimney or flue in a cottage at Thurgoland,
South Yorkshire.   The cottage has been ruined for over thirty
years, and although it was only of one storey, without any chamber
in the roof, the beams themselves are the survivals of a cottage
much older than the one which has been unoccupied for so long, a

Fig. 72.   Ruined cottage at Thurgoland, South Yorkshire, showing the
'bressummer' or fireplace beam, which carried the 'hood' or smoke
chamber above the fire.   The piece cut out of the beam was for a doorway,
by the side of the 'speer' or screen, to the entrance porch.   The fireplace
in the background is a later addition.

cottage with a door by the side of the fireplace, and a 'speer' as a
screen, as is shown by the upright post, and the pointed arch cut in
the beam.   The stone fireplace, and the chimney behind the beam
replaced the original open hearth and hood, and they can hardly be
more than a century old.

Such a great chimney or flue tapered upwards to the roof,
and was generally made of plaster on laths, or more usually on
wattle, or it was formed of less substantial materials, such as cloth
or brown paper, as in some Scotch cottages at the present day.

In the Welsh farm-house, near Strata Florida Abbey, shown in Figs. 35 and 36, the fireplace beam did not run entirely across the house, but was carried by a post at one end. The hood was of plastered wattlework, and at the side was a shelf of the same material, but unplastered, as a roosting-place for the poultry. There was neither stove nor range—simply a hearthstone, and a backstone against the wall[1]. The chimney was not raised above the roof.

It is not surprising that very few of the old hoods now remain. From them were copied the canopied fireplaces which are illustrated so profusely in Mr L. A. Shuffrey's book on *The English Fireplace and its Accessories*. In one example from the north tower of Stokesay Castle, in Shropshire, illustrated by Mr Shuffrey[2], the old wooden tradition was followed in the fireplace beam and its trimmers or cross beams, but all the rest was made of ashlar stone. In the Norman example from Conisborough Castle, in South Yorkshire, which is shown in Fig. 73, the whole fireplace, including the beams, is of stone, and the mason has laboriously copied the straight wooden beams of the carpenter in ashlar, and obtained a level beam by the use of a series of joggled joints. Another fireplace at Conisborough has a longer beam, but is without the stone copies of the wooden trimmers.

In some of the earlier Norman castles of stone, the stone hood is associated with the smoke-hole in the wall. In some examples the hole still remains, a mere slit. Mr Shuffrey describes how it was arranged in connection with the fireplace, and carried through the wall. He points out that one external opening probably 'proved very inefficient for carrying away the smoke during certain conditions of the wind,' and at Castle Hedingham the plan was adopted of making the flue discharge right and left on either side of a flat buttress. Such Norman flues are intermediate in type, though not necessarily in date, between the simple smoke-hole in the wall, and the more developed flue rising vertically in the wall to discharge in a chimney stack.

In a somewhat obscure description of a hall written by Alexander Neckham, in the thirteenth century, the louvre was in the wall and was used as a window ('specularia').

---

[1] There is an illustration of this fireplace in *Archæologia Cambrensis*, XVI (1879), pp. 320–5.

[2] p. 17.

Some cottages in Berkshire, with a combined light- and smoke-hole in the gable, and still in occupation in the last century, have been previously mentioned[1].   In the year 1420 the churchwardens of St Michael's, Bath, whose expenditure upon their buildings has been so often mentioned in this book, paid three pence for making two smoke-holes in the house of William Osborne.   In the same year they paid fourteen pence for taking down the tiles from the

Fig. 73.   Norman fireplace in the Keep of Conisborough Castle, South Yorkshire.   The mason
copied in stone, without change, the wooden
fireplace beam, or bressummer, its trimmers,
and the chimney hood, of the ordinary fireplace
of the period.

house of John Pochin 'in Walcote-strete,' and the payment which follows is three pence for a board bought for one louvre at the same place ('ibidem').   Ten years afterwards a penny was paid for another board for the same purpose at another house.   In the year 1473–4 the churchwardens paid two pence for one 'loverborde.' The seventeenth-century antiquary, Aubrey, wrote that before the Reformation, ordinary men's houses, copyholders and the like,

[1] See p. 249.

'had no chimneys but flues like louvre holes: some of 'em were in being when I was a boy[1].' These references show the confusion which existed as to the meaning of the word louvre, and the difficulty in working out the history of smoke removal from the documents.

Our oldest stone chimneys are of the Norman period, and cylindrical in shape, a form which the writer has seen still in use in stone chimneys in the English Lakes district and in the English speaking part of South Pembrokeshire, which are both remote districts of the North and West respectively. A chimney, of this shape, of ashlar, is more costly in workmanship than is one with straight sides, by which the circular shaft was superseded. It may therefore be reasoned that the cylindrical stone shafts were copied from some material to which the shape was more natural, such as the 'headless barrel,' which served as a chimney in the cottage in Bishop Hall's *Satires*. The history of the chimneys and the fireplaces in wood and in stone cannot be traced without a knowledge of their prototypes, probably in wood or in wattle, and these have generally perished. Sixty years ago, in the oldest cottages of Northumberland, 'a few sticks were twisted together and plastered with clay for a chimney[2],' and we learn that in the year 1808, in the north-eastern district of Bedfordshire 'the cottages and barns, and even some of the farmhouses, are built with wood framework and clay plaster upon a hedgework of splints, which is called wattle-and-dab. The chimnies of the cottages are frequently composed of the same materials[3].'

Mr Ralph Nevill said[4] that he had spoken to workmen in South-West Surrey who had taken down some old chimneys, the shafts of which were formed of wattle and clay. The latter had become as hard as brick. He also says that Rev. G. G. Cooper had informed him of a similar instance along the Sussex border and he quotes a decision of the chief inhabitants and headboroughs of Clare in Suffolk, at their court held on April 17th, 1621, that

[1] 'Antiquarian Repertory' (1678), quoted in *Dictionary of Architectural Publication Society*, *s.v.* Flues.

[2] J. H. Parker, *Domestic Architecture in England from Edward I to Richard II*, p. 200.

[3] T. Batchelor, *General View of the Agriculture of the County of Bedford*, p. 21. He noticed with some astonishment that in the limestone district West of Bedford 'even the chimneys are sometimes built of stone.'

[4] *Old Cottage and Domestic Architecture in South-West Surrey*, p. 19.

every chimney to be erected should be built of brick, 'above the roof of the house fower feete and a halfe upon the paine for every such offence to be hereafter committed the summe of v l[1].' But nearly a century later, in 1719, persons were being fined for their clay chimneys.

Flues were formerly called ' funnels ' and an Act of Parliament of 7 Anne, directs that no timber shall lie nearer than five inches to any chimney funnel or fireplace, and further that all the funnels shall be plastered or pargetted on the inside from the bottom to the top[2]. The size of the flues was settled by another Act of Parliament in 1840. The former Act was the result of the Great Fire of London, the latter of the humanitarianism which concerned itself with the chimney sweepers' boys.

---

[1] *Bury and West Suffolk Archæological Trans.* II, p. 106.
[2] *Dictionary of the Architectural Publication Society, s.v.* Funnel.

CHAPTER XVI

NEW MATERIALS AND CONCLUSION

The changes of the nineteenth century—Cast, wrought, and rolled iron—Steel
in building construction—Reinforced concrete—The uses of lead—Cast
and milled lead—Concrete in foundations—General development of the
methods of construction—Workers and methods in the past and the
present—Conclusion.

Changes in the methods which have been described in the preceding chapters were slow and gradual, until about a century ago.
Long before this time the greater English buildings had felt the
influence of the Renaissance in Italy; but little impression had
been made on the smaller, and their construction had hardly been
influenced. So it was also with the furniture which the buildings
contained, which was marked by a 'sublime indifference to
passing fashions[1].' Mr Arthur Hayden says that bacon cupboard,
linen chest, gate table, ladder-back chair, and windsor chair were
made down to fifty years ago, without departing from the original
patterns of the periods of Charles I and Queen Anne. And,
we may add, it was only the lack of furniture in the ordinary
houses of the Middle Ages which prevented the traditonal patterns
from reaching backwards to a much earlier period than the seventeenth century.

Great changes in building came in with the nineteenth century.
This was due in part to improved means of communication, which
followed steam locomotion, and in part to the greater wealth which
resulted from world-wide trade, and the introduction of machinery
in manufactures. Of more importance still was compulsory elementary education, which brought 'book-learning' to workers
whose previous instruction in the crafts had been verbal, and
derived from tradition. One effect of these changes was such an

[1] Arthur Hayden, *Chats on Cottage and Farmhouse Furniture*, p. 32.

introduction of new materials as had probably never before been experienced in this country. The changes produced by the introduction of new materials are of more importance than those produced by an alteration in the method of using the old, for the character of the work is necessarily conditioned by the nature of the material. It has been stated that 'The Pennant-grit walls of the Roman fort at Gellygaer are precisely similar in their appearance and method of construction to the modern work in the same material in that district: and it would be a difficult task to point out wherein the walls of the Roman houses at Silchester differ from the flint walls of later date. The Roman builders, like the modern, were governed by the material they worked with[1].' So experience, in the course of centuries, has presented certain rules for working which do not alter: thus Joseph Moxon, the seventeenth-century pioneer in building construction books, advised that, in building, no part of the walls should be carried up more than three feet above any other part; he also gave a table of the scantlings of timbers at the several bearings 'for Summers, Girders, Joysts, Rafters, &c., as they are set down in the Act of Parliament for the rebuilding the City of London after the late dreadful Fire: which Scantlins were well consulted by able workmen before they were reduced into an Act[2].'

Iron and steel and concrete are the most important of the new materials of the nineteenth century. The carpenter, who had been for so long a time the principal worker in English building, began to feel the effect of the stream of new materials as long ago as the seventeenth century, but the full flood, which reduced him to the same level as the other craftsmen, has only run during the last century. As Mr Starkie Gardner has picturesquely said: 'In the struggle for existence the smith has continuously encroached on the carpenter's domain. If the smith produced iron chests and bedsteads, iron casements, iron doors and gates, screens and palings, it meant so much work taken from the carpenter. But such things were only the shadows of coming events, and trifling, compared to the inroads of the modern engineer, who has successively annexed shipbuilding, piers, jetties, bridges, roofs, conservatories, railway stations, shop-fronts, temporary halls, churches, huts, and shelters[3].'

---

[1] John Ward, *Romano-British Buildings and Earthworks*, p. 255.
[2] *Mechanick Exercises*, p. 142.
[3] 'Decorative Ironwork' in *The Arts connected with Building*, p. 208.

Iron, like lead, was almost a precious metal until recent times, and was hardly used in ordinary buildings. Even in South Yorkshire, the home of iron workers, its use was restricted to nails and hinges and door handles. The old iron which remains is remarkable for its freedom from rust and for its lasting qualities and durability during periods of time in which modern iron would have decayed away. This is generally held to be due to the smelting of the old iron with charcoal instead of coal or coke.

Wrought iron was used in buildings until the introduction of cast iron in the eighteenth century. The great scale in which the manufacture of iron and steel has been carried on since then has lowered the cost of the material, and it has taken the place of timber for structural purposes.

Sir Robert Smirke, R.A., stated that 'I do not know that I was the earliest to use cast iron girders, but I never saw or heard of any, except at some small factory buildings in Manchester, until about 1810; I was then engaged in rebuilding part of Lord Bathurst's house at Cirencester; for that purpose iron girders between thirty and forty feet long were cast at Coalbrookdale, and the front wall, with other parts of the building, rests upon them[1],' and in the year 1811 a patent was granted to T. Pearsall for cast iron rafters, joists, and skeletons of staircases. The material came into general use, and goods of such a very inferior quality were supplied, that a slight blow would destroy them. This was about the year 1840, and it was felt that it was unsafe to use such a material in building. Rolled iron, which is rolled out from the lump in a mill, took its place between the years 1840 and 1850. The Great Exhibition building of 1851 consisted largely of rolled iron, and the publicity given by such a use led to its more general application[2], but the use of cast iron girders continued for some time longer, and the writer has seen them in buildings erected about the year 1880.

Towards the middle of the nineteenth century, girders and beams built up of rolled iron in various shapes and connected by angle irons were introduced. The most simple and early form of such beams was 'constructed of top and bottom horizontal plates of wrought iron, generally boiler plate connected by angle

---

[1] *Dictionary of the Architectural Publication Society, s.v.* Iron.
[2] *Ibid.*

irons, riveted to a vertical plate in the middle of them.' This form has become standardised, and the horizontal pieces at the top and bottom are known as the top and bottom flange, and the connecting vertical piece is known as the web; since 1866 the joists or girders have been rolled whole with all the plates in one piece. The 'built' up girders have also continued in use.

Changes in building materials now take place rapidly and mild steel has superseded rolled iron for structural purposes; the writer remembers the hesitation with which it was adopted by architects, when it began to take the place of iron, a quarter of a century ago. The forms of the steel joists followed those which had been adopted for rolled iron. In the year 1886 it was said that Bessemer's process for steel, which dates from the year 1858, was 'one which may tend to do away with all mere *cast or wrought iron* in building construction[1],' and the prophecy has come true with respect to purely constructional work.

The attitude of architects to constructional ironwork, half a century ago, when its use was becoming common, may be understood from the following extract from a lecture by Professor George Aitchison at the Royal Institute of British Architects, in February, 1864. He said: 'The use of iron has become so general that it has nearly usurped the place of other materials. In every construction, machine, tool, or article of domestic use, we may find iron doing duty not unfrequently in the most unexpected ways and places.—Architects, therefore, must not alone refuse to profit by the advantages it offers, or, when they are absolutely obliged to use it, give it a sort of grumbling recognition and put it out of sight. Many have used it with success, but by the profession generally it is avoided[2].'

There is as little finality to the varieties of building materials as there is to the methods of construction, or to the styles of architecture. At present the constructor of buildings uses the material which experience has shown to be most suitable *at the price*. There is a choice of metals in building, as there is of styles for architectural design, but the choice is more logically exercised. Lead is used generally for water pipes, often for flashings, frequently for gutters, and sometimes for flats. Wrought iron, the ancient form of the metal, is still used for the old purposes of hinges and grilles, and gates and railings. Its use in England has been

[1] *Dictionary of the Architectural Publication Society*, *s.v.* Steel.   [2] Ibid. *s.v.* Iron.

continuous for more than two thousand years, and smiths with their tools are shown on the English stone carvings of a thousand years ago[1]. Cast iron has found its rightful place, and remains in a restricted use for grates and gutters and pipes, and, occasionally, for more important constructional purposes in columns. Rolled iron is used for fireplace arch bars, but steel is now the constructional metal *par excellence* and its improved manufacture and cheapness have enabled works in metal to be carried out of an importance and size never possible before the nineteenth century.

After iron and steel, lead is now the metal most used in building construction, and its earlier history is interesting. In the days when lead was costly and uncommon, it was chiefly used for covering the roofs of costly buildings. Its decorative value, always associated with costliness in the popular mind, was also appreciated, but plumbing work in connection with water supply and sanitary fittings only came in with the growth of the urban populations in the nineteenth century, on a scale never before seen in this country. This now forms the principal part of the plumber's business and the older uses are quite subordinated to it.

Old lead pipes were jointed, that is they were made out of sheet lead, bent round a core and soldered, and the drawn lead pipes, in general use for water at the present day, were invented in France about a century ago. Before then, although lead pipes were much used in lead-producing districts, and beyond them in the expensive buildings, the water mains were usually made of wood, that universal material for the work of the olden time. Even the upright pipes of pumps and wells were made of bored branches or trunks of trees.

Until the end of the seventeentn century, the lead used in buildings was cast. It was obtained by the plumber in the lump, called the 'pig' or 'sow' and cast into sheets by him, at or near to the building. Now, the lead generally used is rolled into sheets at a mill, and purchased by the plumber in rolls ready for use, an example of that specialisation of industry, which accompanies advancing civilisation, and which has operated in all trades, and not only in those connected with building. The older form is called cast lead, and the modern milled lead.

[1] As on the cross shaft in Halton churchyard near Lancaster which shows Regin forging a sword for Sigurd. Wayland the Smith and his tools are carved on the great cross in Leeds Parish Church.

The casting of lead was done on an oblong table, with raised rims, and the method only varied within narrow limits. The table was first covered with fine damp sand and into this it was possible to press models of letters, numerals or ornaments, which were sometimes arranged in a pattern : these showed in relief on the cast sheet of lead. When the table had been made ready in this way, molten lead was ladled into a receptacle of some kind at one end of the table. Then the lead was poured over the table, and immediately a bar called a 'strike' was rapidly pushed over the table: the ends of the strike were usually rebated and rested on the raised rims of the table. As the strike moved over the table, the surplus molten lead was swept before it, over the end of the table into another receptacle and a sheet of lead was formed. The quantity of lead melted and run was about double that required for the sheet. The strike was raised above the surface of the sand sufficiently to give the required thickness to the sheet of lead, and this height was adjustable, so that the sheet could be cast thinner or thicker as required.

It was difficult to keep the sheet of the same thickness throughout in casting, and flaws were sometimes caused by steam rising from the damp sand. These flaws developed later into holes and cracks in the sheets of lead, when laid upon the roof, and cast lead is responsible for the decay of many old roofs, especially those of churches. It was therefore necessary to use a greater thickness of cast lead than is now used of milled lead, with which a weight as low as 3 lbs. per superficial foot is suitable in some positions. As the cost of lead is conditional on the weight, cast lead is the more expensive, and it has been superseded by milled lead, reduction in cost being a primary cause of changes in building construction. It was not possible to cast lead properly of a weight less than 6 lbs. per superficial foot, and the weights in regular use varied from 7 to 12 lbs., according to the situation ; as an instance, about the year 1735, lead of 9 or 10 lbs. was considered suitable for a flat[1].

At some large buildings, originally covered with cast lead, the old method of casting is still retained for repairs to the roofs. Such are St Paul's Cathedral, London, and Wentworth Woodhouse, near Sheffield: at the former, according to Mr F. W. Troup, the

---

[1] *Dictionary of the Architectural Publication Society, s.v.* Lead.

sheets are cast 'on a very large casting frame 17 feet long by some 16 feet wide[1].'

Milled lead was invented during the last quarter of the seventeenth century, that time of change, when the modern period began in building construction. Early milled lead was rolled in lengths of 34 feet, which was twice the length of the largest casting table, but the manufacturers made the easily understood mistake of rolling it too thin. At Greenwich Hospital, which had been covered with milled lead, 'it rained in almost all over the hospital,' in the year 1700, from which it would appear that milled lead was used by Sir Christopher Wren. 'Parliament sent the master and wardens of the Plumbers' Company to view it, and they unanimously declared that milled lead was not fit to be used' on buildings[2]. The prejudice against it lingered on until the nineteenth century.

Lead seems to have always been laid upon boarding, as it is at present, and in the Middle Ages in the North such boards were known as sarking boards, and the word has survived in the dialect of Northumberland. The word 'sark' is an Old English word for shirt, and is used with that meaning in *Beowulf*, so that sarking was that which clothed the building like an under-garment, and was an apposite simile of the early builders.

The favourite joint or seam between laid sheets of lead was formed by wrapping the edges of the two sheets together into a kind of semicircle, and in the year 1736 the one piece of lead was called 'orlop' (probably o'erlap), and the other the 'stander,' and the tools used to make the joint were a 'dresser' and a 'seaming mallet.'

Concrete is another material of which the use on a large scale is a modern development. Although we have seen that various forms which approached it were used in the past, such as walls roughly built of mortar and small rubble, and 'lime ash' used in floors, the use of concrete proper only arose with the increase of large engineering works during the nineteenth century.

Dance, in the year 1770 'having to sink the principal foundations of Newgate prison to a depth of 40 ft., in consequence of their site being partly on the ancient ditch of London wall, threw into the bog cartloads of whole and broken bricks and cartloads of

[1] 'External Leadwork' in *The Arts Connected with Building*, p. 191. *Dictionary of the Architectural Publication Society*, s.v. Lead.

mortar in the proportion of one to four[1].' This made a good foundation, but was not concrete in the modern sense.

Sir Robert Smirke, an architect of the early nineteenth century, used concrete on a large scale for the foundations of the General Post Office, in St Martin's-le-Grand. He covered the whole site, where 'a greater diversity of subsoil was never before exposed to view[2].' But even this concrete was not mixed as it is to-day; Smirke had the stones or other materials rammed in place, grouting each layer with mortar[1]. Concrete is now mixed before it is placed in position. The name was given between 1820 and 1830.

The first cement was made from natural nodules found about the mouth of the Thames, and was known as Roman cement. But the great impetus to concrete work was given by the discovery that cement, which had previously been supposed to be purely natural, could be artificially made by burning a mixture of the necessary materials in the due proportion. Such was Portland cement, and its manufacture has now been brought to a great pitch of perfection, and has enormously increased the possible uses of concrete (see also p. 147).

The most remarkable modern development in the use of iron and steel in building has been in its combination with concrete. After the introduction of the latter material it was soon found that it could be strengthened in floors by the use of iron, and afterwards steel, joists and that the assistance was mutual. But its use was of a very elementary kind, until the beginning of the present century, when the strengthening of concrete by metal work, on a large scale, was adopted in France and afterwards in this country. This is called 'reinforced concrete,' and so important has it become, that the description of the various systems occupies a large space in every modern book of building construction.

The increase in wealth and the relative decrease in the cost of manufacture in modern times, have brought into common use many other materials, which, like lead, were previously restricted to the service of the wealthy. Glass is a material of this kind, for its regular use for the admission of light into ordinary buildings is comparatively recent.

Crown glass was the 'common description of window glass

[1] *Dictionary of the Architectural Publication Society*, *s.v.* Concrete.
[2] James Elmes, Surveyor of the Port of London, in *Notes and Queries*, 2nd Series, VI, p. 290.

used in England almost exclusively before the abolition of the excise duty.' It is greenish in colour, not regular in appearance, and wavy in surface. Crown glass was blown into the shape of a flattened sphere, and spun into a flattened circular plate, called a 'table,' of which twenty-four formed a 'crate.' A lump was left in the sheet where the glass had been attached to the pipe: this is the 'bullion,' 'bull's eye,' or 'bottle end,' which is still common in the windows of old cottages, where the cheapest kind of glass was used. Although the bottle ends are the cheapest and most inferior part of the sheet, they became fashionable in the so-called Queen Anne revival, and were bought from old cottages for re-use in good modern houses.

The glass making industry was sufficiently prosperous in 1746 to tempt the government to impose an excise duty, which is said to have been so onerous that it is a matter for wonder that the manufacture survived[1]. This continued for nearly ninety years, and on its repeal the glass industry greatly increased; crown glass was superseded by sheet glass, which is technically, if not æsthetically, superior to crown glass.

Plate glass is the most technically excellent kind of glass for building use, and is of French origin. According to the *Dictionary of the Architectural Publication Society* the first plate glass works in England were established in 1773 at Ravenhead, near Prescot, Lancashire, by the 'Governor and Company of British Plate Glass Makers,' under a charter[2].

Coloured glass, whether stained or painted, was never used in such ordinary buildings as are described in this book. It was usually arranged in a pictorial manner, great interest in so-called

---

[1] *Encyclopaedia Britannica*, 11th edition, XII, p. 104.

[2] *Dictionary of the Architectural Publication Society*, s.vv. Glass and Plate glass. According to Robert Brown (*Domestic Architecture*, p. 222), in England 'the first plate glass was made in 1673, at Lambeth, and its manufacture was introduced by the Duke of Buckingham, who, for that purpose, brought over several Venetian work-man.' The Houghton letters of the end of the seventeenth century state that plate glass was manufactured at two places in this country. But the names of the kinds of glass have varied, as have the kinds themselves, and plate glass is not mentioned by that name in a 'Builder's Dictionary,' in an anonymous *Rudiments of Architecture*, of which the 3rd edition was published at Dundee in 1799. It says of glass 'there are many sorts; as crown glass, French or Normandy glass, German glass, Newcastle glass, Bristol glass, looking glass, and jealous glass; which last is of that nature that it cannot be seen thro', yet it admits of the light thro' it.' This 'Dictionary' was copied from that of Hoppus in his 5th edition (1755) of Salmon's 'Palladio Londinensis,' which in its turn had been copied from earlier dictionaries, such as those of Neve and Moxon.

'stained glass windows' has caused far more to be written of it, than of all the other kinds of building glass together.

Paint is also a material of which the use in ordinary buildings is comparatively recent. Although buildings had been painted before the modern period, which began in the latter half of the seventeenth century, it was often with strange materials such as the bullocks' blood before mentioned[1], or with an extract of liver-work called archil[2]. The use of oil paint for exteriors apparently began in the seventeenth century and Sir Balthazar Gerbier implied that nut oil was used as he said that the carpenters, 'being of an honest Joseph's profession are as deserving to be well paid as the Painters, who do but spend the sweat of Wallnuts (to wit oyl), the Carpenters that of their brows[3].'

The principal materials used in ordinary buildings have now been considered in detail, and it is possible to take a wider view of the subject. The true value of history is that it throws some light not only on the past, with which we have before been occupied, but, in some degree on the future also, and as progress in building construction has continued, century after century, in a certain direction of development, it may reasonably be assumed that in the future it will follow along the path of the past.

The general impression received from the study of the development of building is that construction ever tends to become more and more complicated : the parts of the framework become more numerous, and the joints themselves become more complicated, and there is also an increase in the number and the variety of the materials used, as we have just seen.

Ordinary buildings also tend to be built better, and hence, more permanently, in spite of the generally accepted opinion to the contrary. Thatch is ousted by slates, and stone and brickwork take the place of lath-and-plaster, and of wattle-and-daub. This is due to the general growth of civilisation, and, to use a homely expression, may be placed to the credit side of the account. As has been stated before when comparing the old and the new, many of the solidly built 'old cottages' were probably not built for use as cottages, but have descended in the social scale in the course of time.

[1] Ante, p. 159.
[2] S. O. Addy, *Evolution of the English House*, pp. 71, 91, 123.
[3] *Counsel and Advice to all Builders*, p. 48.

On the debit side of the account is to be placed the increasing
ugliness of ordinary, or as it may be called, natural building,
because the culture of the present day has passed the stage when
Art is natural and spontaneous, coming as it were hand-in-hand
with the work.   In the end, perfection in technique kills Art; to
attain to it some difficulty with the material is necessary, and
further, there should be individual effort and interest on the part
of the workman, who must feel some pleasure in his work.   John
Bunyan, the Puritan author of *The Pilgrim's Progress*, has well
expressed this necessary egotism of the artist in his rhymed
'Apology' for his work.   He says:

> But yet I did not think
> To show to all the world my pen and ink
> In such a mode: I only thought to make
> I knew not what: nor did I undertake
> Thereby to please my neighbour: no, not I:
> I did it mine own self to gratify.

That is impossible with ordinary work in a modern factory, or
even in the construction of a modern building.   Artistic work
cannot be attained under the conditions in which most building
work is now carried on.   Civilisation has moved forward on the
road of Progress, and in so doing has left Art forlorn by the
wayside.

The old methods of craftsmanship are vanishing with the
changed conditions of education and industry, and it is a matter
for regret that they cannot be adequately described in writing.   In
this book it has only been possible to study the construction, for
craftsmanship is often a matter of mere sleight of hand.   It may
be understood, but can hardly be described, just as the technicalities
of fine architectural draughtsmanship cannot be learned from books.

Other movements of to-day are in a growing tendency to
specialisation, which has already been noticed, and an increase of
routine in work, which tend to make the work of less interest to
the worker, and interest in work and intelligence in its application
seem to go together.   There has also been a permanent trend to
faster working.   The time which was taken to build the old
buildings of the past is astonishing to us in the twentieth century.
There is a tradition of the great timber barn at Gunthwaite, South
Yorkshire, that 'one of the apprentice lads of the carpenter who
built the barn was employed during the whole of his apprenticeship

in making the pegs required[1]' for the joints of the timberwork. Faster work is one of the accompaniments of progress in civilisation and against it the workers are always struggling.

The social position of the worker has also greatly improved during the historic period, for a thousand years ago the building workman was a slave, a chattel, who was transferred from one owner to another, with the land on which he worked : to-day his descendants not only vote for their rulers, but may themselves join the ruling class by entering Parliament[2].

During the nineteenth century the woman-worker ceased to be employed in the building trades. And previously women seem to have been occupied only in unskilled work, such as the carrying of the straw to the thatcher on the roof[3], or the treading-in of the straw in daub[4], or the gathering of moss for the roofing slates[5]. *The Records of the Borough of Leicester* show that in 1327–8 a woman was paid 1*d.* per day for collecting stones out of the water in connection with work at the North Bridge, which reminds us of Margery Daw in the nursery rhyme, who was to have only 1*d.* a day because she could not work any faster. The *Records* also show that in the financial period 1377–9 two women worked for three weeks as labourers to the workman who made the wattle-and-daub coping (or possibly parapet) of the town wall. Women therefore appear to have acted as 'labourers' and the writer knows of no instance of a woman-mason or a woman-carpenter. The better-paid work seems to have been always reserved by the men for themselves.

The movement of the stream of cultural progress, although usually slow, is certainly onward, and the building trade is a part of that stream, so that at any period the status of the building worker, accords on the whole, with that of the time. The country has become more wealthy to an immense extent, and through the centuries the builder seems to have had his share in the general prosperity. A careful examination of the old buildings in the

---

[1] *Sheffield Daily Telegraph,* August 3, 1895. The barn is shown in Fig. 41.

[2] Some pessimists may wonder whether the shadow has not been mistaken for the substance. Gerbier (*Counsel and Advice to all Builders*) advised that the scaffolding should be ready for the workmen at the beginning of the day, or they would go off for a day's enjoyment, and during the present war, workmen from the old staple trades of Sheffield, who have obtained employment in the comparatively modern armament factories of the city, have been dissatisfied with the advanced organisation and discipline which they compared unfavourably with the methods of their own older trades.

[3] Ante p. 194.   [4] Ante p. 142.   [5] Ante p. 180.

district about Sheffield shows that this prosperity in the building trade ran in cycles of about three generations. At present there is a decline from a recent period of prosperity[1].

On a border of Sherwood Forest an old coach road, now deep in sand, but once frequented by Dick Turpin, curves round to avoid a hollow in which, sheltered from the weather, stands an old cottage. From the hollow the old road runs across a common, which in July has the rich purple flowers of the heath set among its green, and behind all, forming a background to road and common and cottage, there stretches, to right and left, the dark foliage of the ancient forest. The cottage stands in its own little enclosure : its walls are of brick, its roof of tiles, both of the beautiful red colour which is possessed by the old Nottinghamshire tiles and bricks. Herbs for use and flowers for display border the path to the door, and a wooden water butt stands against the house. Grey-brown pine cones, rustic weather glasses, are hung with string to nails driven into the joints of the red brickwork of the end gable, and swing gently with the breeze which also stirs the forest leaves. The cottage and its enclosure seem to belong to the landscape in which they are set. A closer examination shows that the cottage itself is a patchwork of time, for half-hidden in the brickwork there are timber posts, which have remained from a time when the cottage walls were of some material more simple than burnt brick. These original walls have gone, but the principal oaken framework has remained, and to it the brick walls are a later addition, just as the old coach road, and the cottage itself, were once added to the old landscape of wood and heath. To-day, in an industrial country, such a landscape and such a cottage are each somewhat of an anachronism, for the landscape of forest and common is that of an earlier England, and the cottage is a survival from a more simple culture.

There is a tendency to forget that ordinary buildings, as well as the methods of their construction, usually are in accord with the general state of civilisation at the time. So that reproductions of old styles of architecture, and of old methods of buildings, are necessarily out of place, unless the old surroundings are reproduced also. Buildings cannot be satisfactorily designed apart from their surroundings, which are not merely other buildings and the

[1] This was written before the war, which will accentuate the decline.

landscape in its widest meaning, but also the general culture of the period.   Such a building as the old cottage of the forest border, at the time of its erection, was a natural expression of the spirit of the time in concrete materials, and if such a cottage be built to-day, it will only be out of place unless there can also be recalled from the past the old life and the old conditions, with the tinkle of the leper's bell in the forest, or the ringing challenge of the highwayman on the heath.

The value of old buildings as works of art does not lie so much in their suitability for reproduction, as in their power for inspiration, in the intangible principles which were given expression in the different materials and workmanship, whose story in England has been partly told in this book.

INDEX